NATURAL ENVIRONMENT RESEARCH COUNCIL

INSTITUTE OF GEOLOGICAL SCIENCES

Report No. 70/17

Geochemistry of Recent sediments from the central north-eastern Irish Sea

D. S. Cronan, B.Sc., Ph.D., D.I.C.

London: Her Majesty's Stationery Office 1970

THE INSTITUTE OF GEOLOGICAL SCIENCES

was formed by the incorporation of the

GEOLOGICAL SURVEY OF GREAT BRITAIN

and the

MUSEUM OF PRACTICAL GEOLOGY

with

OVERSEAS GEOLOGICAL SURVEYS

and is a constituent body of the

NATURAL ENVIRONMENT RESEARCH COUNCIL

SBN 11 8801316 8

Reference to this report

It is recommended that reference to this work be made in the following form:

Cronan, D.S. 1970. Geochemistry of Recent sediments from the central north-eastern Irish Sea. *Rep. No. 70/17, Inst. Geol. Sci. 20 pp.*

CONTENTS

SUMMARY

Determinations of B, Ba, Cr, Fe, Ga, K, Mn, P, Sr, Ti, Zr and $CaCO_3$ in 115 samples of sea-bed sediments and 25 clay concentrates from the sediment surface in the central north-eastern Irish Sea, together with Co, Cu, Ni, V and Zn in 12 samples, have shown the sediments to vary markedly in composition both with grain size and mineralogical nature. The carbonate fraction of the sediments consists largely of skeletal remains, and its distribution throughout the basin is thought to reflect the energy level of the environment in that it is most abundant in the west where strong tidal currents prevent the accumulation of clastic detritus and thus the burial of the shells, and least abundant in the east where weak currents allow the deposition of large quantities of mud. In this fraction is concentrated most of the Sr in the sediments, together with lesser amounts of Mn and other elements. The non-carbonate fraction of the sediments is thought to originate principally from the reworking and grain size fractionation by tidal currents of Pleistocene glacial deposits on the floor of the Irish Sea, and its compositional variations are considered largely to reflect compositional differences between the various grain size classes of the parent glacial tills with some modifications resulting from the selective breakdown of unstable minerals and adsorption of elements from seawater. In this fraction, and especially in the clays, are concentrated major portions of B, Ba, Cr, Fe, Ga, K, Mn, P and Ti, while Zr is most abundant in the muddy sands. Adjacent to the English coast, certain anomalies in the composition of the sediments are inconsistent with some of their constituents having been derived from reworked glacial debris within the Irish Sea basin and suggest that these constituents may have been derived from the English mainland.

SOMMAIRE

Des déterminations de B, Ba, Cr, Fe, Ga, K, Mn, P, Sr, Ti, Zr et de $CaCO_3$ dans 115 échantillons de sédiments marins, et de 25 concentrés argileux de la Mer Irlandaise du nord-est central, avec celles de Co, Cu, Ni, V et de Zn dans 12 échantillons, ont montré de grandes variations de compositions et en grandeur des grains et en nature minéralogique. Le carbonate, qui forme partie des sédiments, consiste principalement en restes squelettiques, dont la distribution par tout le bassin réfléchiraient, semble-t-il, le niveau d'énergie du milieu, puisqu'il abonde à l'ouest, où des courants forts empêchent l'accumulation de matérial clastique de sorte que les coquillages ne sont pas couverts. À l'est, cependant, il est plus rare puisque les courants faibles n'y permettent pas le dépôt de beaucoup de boue. Dans cette partie se trouve concentrée la plupart du Sr dans les sédiments, avec moins de Mn et d'autres éléments. Les sédiments, qui ne sont pas carbonates, pourraient provenir principalement des dépôts glaciaires Pleistocènes au fond de la Mer Irlandaise où les grains ont été remainiés et fragmentés par l'action des courants de la marée. Des variations en composition semblent réfléter des différences en composition entre les grains variés des moraines avec des modifications produites par la désintégration sélective de minéraux instables et par l'adsorption d'éléments de l'eau de mer. Dans cette partie, et surtout dans les argiles, sont concentrées des quantités majeures de B, Ba, Cr, Fe, Ga, K, Mn, P et de Ti, tandis que Zr est plus abondant dans les sables boueux. Près du littoral d'Angleterre la composition des sédiments montre des anomalies qui suggèrent que quelques-uns des composants ne proviennent pas de détritus glaciaire remanié dans le bassin de la Mer Irlandaise, mais qu'ils ont pu venir de l'Angleterre.

ZUSAMMENFASSUNG

Bestimmungen von B, Ba, Cr, Fe, Ga, K, Mn, P, Sr, Ti, Zr und $CaCO_3$ in 115 Proben von Meeresgrund-sedimente und 25 Ton-Konzentrate aus der Sediment-Oberfläche in der Zentral-Nordöstlichen Irischer See zusammen mit Co, Cu, Ni, V und Zn in 12 Proben zeigten, dass die Zusammensetzung der Sedimente in Korngrösse und mineralogischer Natur bedeutend verschieden ist. Der Karbonatteil der Sedimente besteht grossenteils aus Skelettresten: und es ist gehalten, dass die Verteilung des Karbonatteiles ganz durch den Becken die Energiehöhe des Milieus reflektiert, weil sie am höchsten in Westen ist, wo starke Gezeiten-strömungen Anhäufung der klastischen Zerfallstoffe und folglich die Begrabung der Muscheln hindern, und am niedrigsten im Osten, wo schwache Gezeitenströmungen die Ablagerung von grossen Quantitäten von Schlamm erlauben. In diesem Teil konzentriert sich die grösste Quantität vom Sr in den Sedimenten, zusa-mmen mit kleineren Quantitäten von Mn und anderen Elementen. Es ist gehalten, dass der Nicht-Karbonat Teil der Sedimente aus dem Aufarbeiten und Korngrössenbrechen durch Gezeitenströmungen von zum Pleis-tozän gehörigen Gletscherablagerungen auf dem Boden der Irischen See hauptsächlich entsteht. Es ist gehalten, dass seine Mischungsverschiedenheiten Mischungsveränderungen zwischen den mannigfaltigen Korngrössenklassen der Urgletscherblocklehm grossenteils reflektieren. Es gibt einige Modifikationen, die aus selektiver Auflösung von unbeständigen Mineralien und Adsorption von Elementen aus Seewasser folgen. In diesem Teil, und hauptsächlich in den Tonen konzentrieren sich grössere Teile von B, Ba, Cr, Fe, Ga, K, Mn, P und Ti, wogegen Zr in den schlammigen Sänden am reichsten ist. Neben der englischen Küste sind gewisse Anomalien in der Zusammensetzung der Sedimente mit einigen von ihren Bestandteilen nicht zusammenhangend, in dem sie von aufgearbeiteten Gletschertrümmern im Irischen Bechen abgeleitet sind. Von diesen Anomalien scheint es möglich, dass diese Bestandteile vom englischen Festland abgeleitet sind.

Geochemistry of Recent sediments from the central north-eastern Irish Sea

D.S. CRONAN, B.Sc., Ph.D., D.I.C.

INTRODUCTION

This paper describes certain aspects of the geochemistry of a suite of shallow water surface sediments collected by the Institute of Geological Sciences in 1967 from the central part of the north-eastern Irish Sea (Fig. 1). The samples were obtained with a Shipek grab and the stations were located on a grid (Fig. 1), positions being fixed with reference to the North British (3B) Decca Navigational Chain.

A total of 115 sediment samples have been analysed for a variety of elements, together with clay mineral determinations on selected clay concentrates. The textural properties and origin of these sediments have been discussed in a previous paper (Cronan 1969) but will be briefly summarised here. They are mixtures of gravel, sand and mud (Fig. 2) containing both terrigenous and non-terrigenous constituents in variable proportions. Over much of the western portion of the area they consist largely of shell detritus mixed with small pebbles and coarse sands, the shell material ranging in concentration up to almost 80 per cent of the deposits. Towards the centre of the basin the sediments become finer, grading through extensive deposits of shelly, slightly gravelly sands into well sorted medium and fine yellow sands, which in turn grade eastwards into grey sandy muds and black muds about 10 to 15 miles off the Cumberland coast. Nearer the English mainland the deposits become coarser, clean, well-sorted sands occurring to the north-west of Morecambe Bay and muddy gravels to the south. This distribution pattern is considered to result largely from the reworking of glacial deposits on the floor of the Irish Sea and their redistribution under the influence of tidal current activity. Tidal current data and sediment colours suggest oxidising conditions over most of the basin, but possible reducing conditions off the Cumberland coast where the black mud is accumulating.

ANALYTICAL METHODS

Immediately following collection, each sample was sealed in a glass jar together with the seawater in the grab bucket. After transport to the laboratory, the samples were allowed to stand until suspended material had settled out. Each sample was then sieved through an 8 mesh B.S. sieve to remove particles greater than 2 mm in diameter and the remaining material was homogenised and sub-sampled for analysis. Representative sub-samples weighing approximately 50 g were sent to the Geochemical Division of the Institute for chemical analysis.

The analyses were conducted using a variety of standard techniques. Optical spectrographic methods were used in attempting to determine B, Ba, Co, Cr, Ga, Fe, Mn, Mo, Ni, Sn, Sr, Ti, V and Zr, and atomic absorption spectrophotometry for Cu and Zn. Phosphorus was determined absorptiometrically as phospho-vanado-molybdic acid, K by flame photometry and carbon dioxide by acid evolution and absorption in sofnolite and magnesium perchlorate. Precision depends on the analytical method used, but is better than $\pm$ 15-20 per cent.

Twenty five samples were sub-sampled for clay mineral analysis. These were carefully disaggregated and sieved wet through a 240 mesh B.S. sieve. The suspensions were transfered to glass cylinders and allowed to settle. After an appropriate time, aliquots containing $< 2\ \mu$ clay were withdrawn from each suspension and evaporated to dryness at room temperature. Portions of these were sent to the Geochemical Division of the Institute and the remainder to the Petrographical Department for X-ray diffraction analysis.

CLAY MINERALOGY

The samples selected for clay mineral analysis were all obtained from the area of muddy sediments in the eastern half of the basin. Quartz, calcite, illite group minerals, kaolinite and a mineral or minerals having a line at 14 Å were found to be present in all samples. Moreover, very little variation in the proportions of these minerals could be detected throughout the area of clay deposition. Six specimens were re-examined after saturation with glycerol in order to establish the presence or absence of minerals of the smectite group. In only one sample did the 14Å line expand to 17Å confirming the presence of a smectite. In the other 5 samples the position of the 14 Å line was not affected by glycerol saturation suggesting the presence of minerals of the chlorite group. Heat treatment of 4 of the 5 samples confirmed the presence of chlorite. Average proportions of the minerals present in the clays are given below.

	Percentages
Illite Group Minerals	> 75
Kaolinite	< 5
14Å Minerals	< 5
Calcite and quartz	< 15

TABLE 1. Mineralogical composition of the clay fraction.

The clay mineral assemblage is fairly uniform in composition throughout the area of mud deposition in the central north-eastern Irish Sea, and is dominated by members of the illite (clay-mica) group of minerals with subordinate concentrations of chlorite and kaolin-

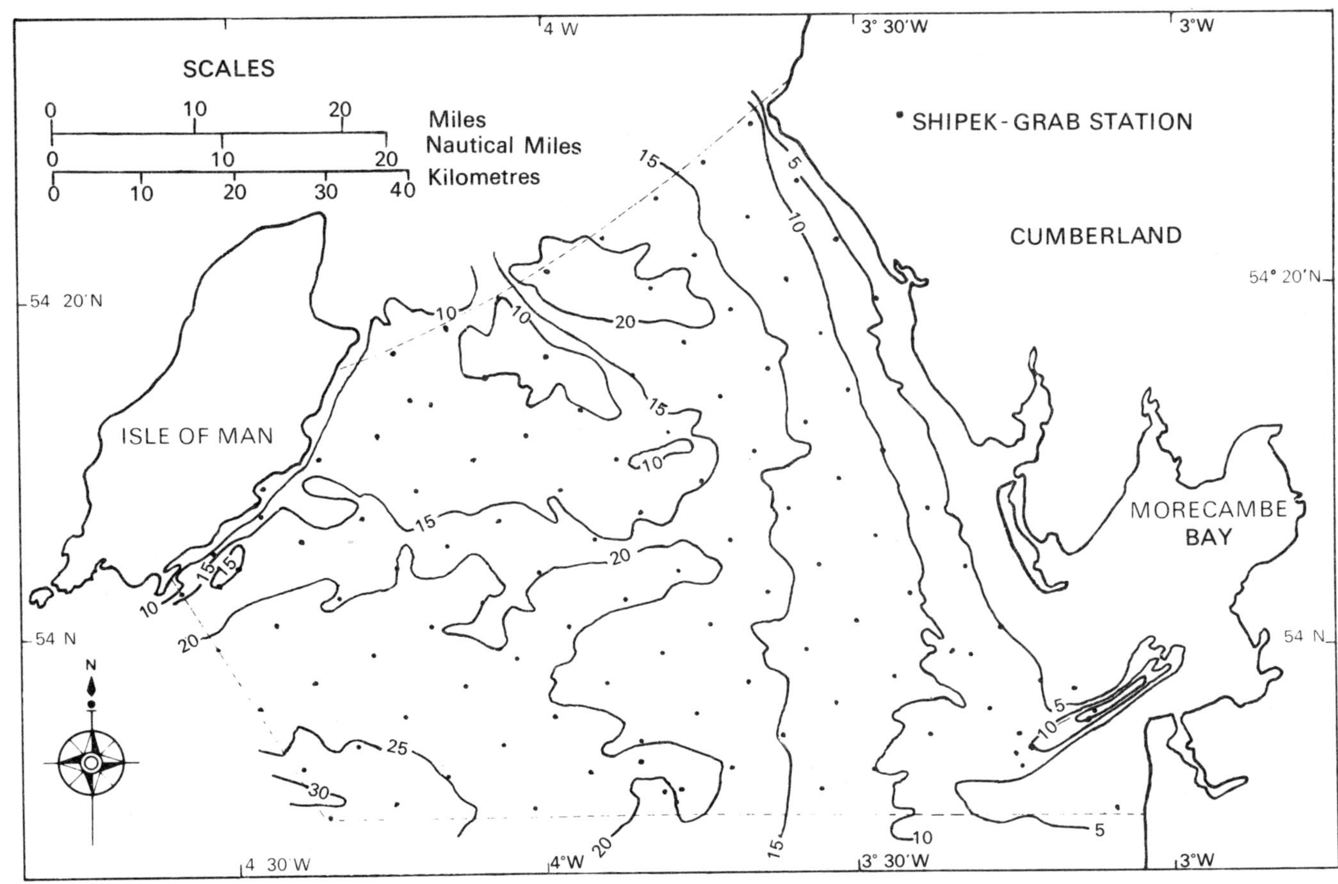

Fig. 1 - Location of the survey area and sampling stations.

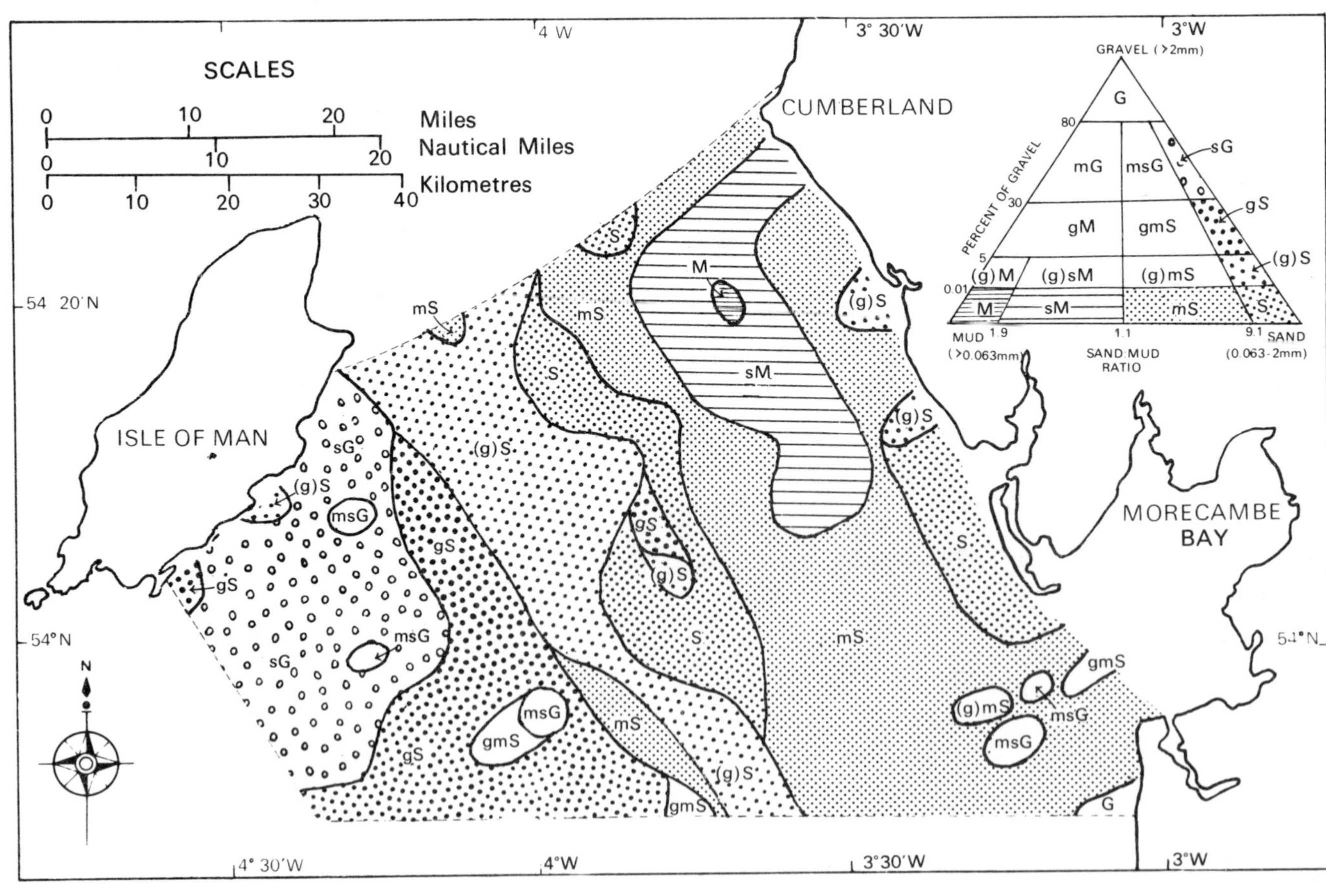

Fig. 2 - Distribution of sediments classified according to the scheme of Folk (1954).

ite. According to Griffin and others (1968), there is no evidence for the in-situ formation of illite in the marine environment, and its abundance in marine sediments is considered to reflect supply from continental sources. This conclusion is supported by the great ages shown by marine illites using potassium – argon geochronological techniques. Hurley and others (1963) found ages of hundreds of million of years for illites in recently deposited marine sediments indicating their derivation from ancient continental rocks. Biscaye (1965) found illite to be the most abundant mineral in the $< 2\ \mu$ fraction of Atlantic sediments and considers this to reflect the high concentration of micas in many rock types on the surrounding land areas, the widespread abundance of illite in many soils, and its relative resistance to chemical weathering. In addition, Berry and Johns (1966) found illite to be ubiquitous in sediments from the North Atlantic and Arctic Oceans and considered that, together with other clay minerals present in the sediments, it had been transported to the marine environment from terrestrial sources by ice rafting and ocean currents.

In view of these considerations, the clay mineral assemblage in the central north-eastern Irish Sea is likely to be detrital in origin and reflect the mineral composition of the source rocks from which the sediments have been derived. On the basis of tidal current data and textural considerations, Cronan (1969) concluded that most of the sediments in the central north-eastern Irish Sea basin have been derived from the reworking of glacial debris on the floor of the Irish Sea itself. The nature of the clay mineral assemblage in the sediments is not at variance with this hypothesis. Frye and others (1963) noted that glacial tills in Illinois which had been derived from lower Palaeozoic rocks were enriched in illite, and Allen and Johns (1960) found illite to be abundant in glacial tills from eastern North America where outcrops of lower Palaeozoic and Pre-Cambrian rocks are numerous. North-west European glacial tills contain material derived largely from the Pre-Cambrian and Palaeozoic formations of Scandinavia, the Baltic Shield and the northern United Kingdom in which mica and illite-rich rocks such as mica-schist and gneiss, shale, slate and greywacke abound. Wilklander and others (1956) noted that illite was commonly the main clay constituent of Swedish soils, and Glentworth and others (1964), Goodyear (1962) and Perrin (1957) found it to be the most abundant clay mineral in glacial deposits from various parts of the United Kingdom. Thus the preponderance of illite in the central north-eastern Irish Sea muds is in accord with what might be expected on the basis of the likely composition of the rocks from which the muds have ultimately been derived. The lack of variation in the composition of the clay mineral assemblage within the basin suggests a single source for the sediments, but does not preclude the local derivation of a proportion of the materials from the English mainland by coastal erosion or river discharge. However, if there is a significant contribution into the basin by these processes, the material from which the clay minerals have been derived is likely to be very similar in composition to the glacial deposits presently being reworked on the floor of the Irish Sea.

CALCIUM CARBONATE

On the basis of the carbon dioxide analyses, the distribution of carbonate materials in the central north-eastern Irish Sea is charted in Figure 3, the results being expressed as percentages of calcium carbonate. Most of the carbonate is in the form of biogenic skeletal debris such as lamellibranchs and gastropods together with subordinate amounts of algal, bryozoan and other organic remains. This material varies in size from whole shells up to three or four centimetres across, down to finely comminuted debris.

Figure 3 shows that the highest concentrations of calcium carbonate occur in the west of the area, and, in particular, just off the east coast of the Isle of Man. Here the carbonate is present mainly in the form of coarse whole and broken mollusc shells and algal fragments, although at one locality just off Langness Point there are deposits consisting almost entirely of broken shells in the coarse sand to fine gravel size range.

The carbonate fraction of the sediments decreases towards the east of the area. Concentrations of between 20 and 40 per cent $CaCO_3$ occur over much of the west-central part, dropping to below 20 per cent in the centre and to below 10 per cent in the east. Carbonates locally increase in amount off Morecambe Bay, but are present in concentrations of less than 10 per cent along the rest of the English coastal strip.

In general, the distribution of carbonate material follows the overall grain size variation of the sediments. Calculation of the correlation coefficient between mean size and total $CaCO_3$ gave a value of $+0.53$, significant at the 0.01 level. Carbonates are most abundant in the coarse sandy gravels from the west and decrease in concentration as the mean size of the sediments decreases, reaching a minimum in the area of sandy mud deposition off the Cumberland coast. The relationship between carbonate concentration and the grain-size of the sediments probably largely reflects the energy level of the environment of deposition. On the basis of tidal current data, Cronan (1969) considered the west of the basin to be a high energy environment with little or no detrital sedimentation and thus in this area calcareous skeletal material would remain exposed after accumulation on the sea floor. By contrast, the accumulation of detrital sediments under the progressively lower energy conditions in the centre and east of the basin is likely to reduce the abundance of calcareous organisms at the surface in these areas by burial. In the east, whole shells are rare in the sediments, and the little carbonate material present is largely in the form of fine shell debris and

	B (ppm)	Ba (ppm)	Cr (ppm)	Fe (°/o)	Ga (ppm)	K (°/o)	Mn (ppm)	P (ppm)	Sr (ppm)	Ti (ppm)	Zr (ppm)	Co (ppm)	Cu (ppm)	Ni (ppm)	Mo (ppm)	Sn (ppm)	V (ppm)	Zn (ppm)
1	32	252	17	1.59	n.d.	0.99	637	439	302	1021	168	3	7	18	n.d.	n.d.	29	44
2	140	568	56	6.33	8.3	2.94	2720	1316	111	2160	48							
3			89	3.72			1401			3588		27	137	37	20		200	200
4	115		85	1.73		2.11	540	780		5200		10	106	38			152	
5	67	319	53	3.95	12.9	1.33	860	506	150	3548	307	9	11.6	22			95	
6			102	5.06			4784	1052		4548		101	323	211	18		215	
7	230	2300	90	6.50	20	2.50	6700	1500	180	4600	150	74	250	225	27	1.5	120	165
8	10	425	100	5.63	15	2.09	950	1050	375	5700	165	25	55	75	1.5	2.0	135	70

1. Average composition of 115 surface sediments from the central north-eastern Irish Sea (Co, Cu, Ni, Mo, Sn, V, Zn, 12 samples).
2. Average composition of 25 clays from the central north-eastern Irish Sea.
3. Average composition of 14 Antarctic pelagic sediment cores (Angino 1966).
4. Average composition of 10 near shore Arctic sediment samples (Macdougall and Harriss 1969).
5. Average composition of 27 sediment samples from the Gulf of Paria (Hirst 1962).
6. Average composition of 183 Pacific pelagic clays (Cronan 1969a)
7. Average composition of deep-sea clays (Turekian and Wedephol 1961).
8. Crustal abundance (Taylor 1964).

TABLE 2 - Average composition of modern marine sediments from various environments.

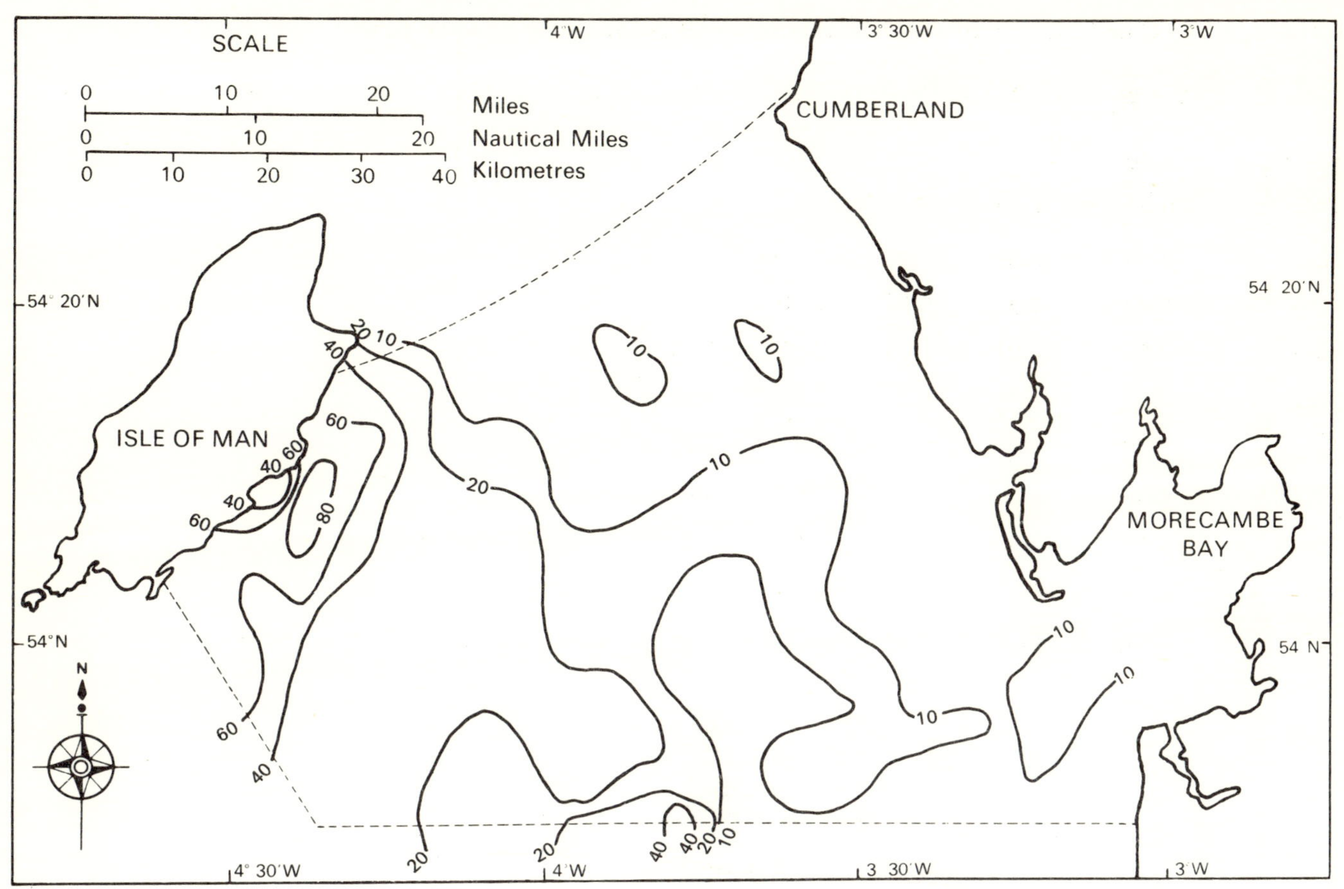

Fig. 3 - Distribution of calcium carbonate (in weight per cent).

calcite grains. A large proportion of these have probably been transported to their present site of deposition by tidal currents, and their concentration in the sediments is being effectively diluted by the large amounts of non-carbonate materials also being deposited. The slight increase in calcium carbonate off Morecambe Bay might reflect the increased tidal current velocities in this area restricting the accumulation of detrital sediments.

ELEMENT ABUNDANCES

Using the techniques outlined above the 115 bulk sediment samples, and the 25 separated clay fractions, were analysed for the elements B, Ba, Cr, Fe, Ga, K, Mn, P, Sr, Ti and Zr. In addition, 12 of the bulk samples from widely separated localities were further analysed for Co, Cu, Ni, Mo, Sn, V and Zn. Detection limits for certain of the elements were as follows: B, 10 ppm; Ba, 30 ppm; Co, 5 ppm; Cr, 10 ppm; Ga, 10 ppm; Mo, 10 ppm; Sn, 10 ppm; Sr, 30 ppm; V, 30 ppm and Zr, 30 ppm. When an element was beneath detection in any particular sample its concentration was taken to be half the detection limit for the computation of averages. Mean concentrations of the elements determined in the bulk samples are presented in Table 2, together with the averages for the clays and some data on the average composition of recent sediments from other areas.

ELEMENT ASSOCIATIONS

Inter-element Relationships

A matrix of correlation coefficients for all the constituents determined in the sediments is presented in Table 3. Most elements are positively correlated with each other and negatively correlated with calcium carbonate, although Mn and Sr are positively correlated both with each other and with calcium carbonate. These observations can be partly explained in terms of a model proposed by Chave and Mackenzie (1961). These authors suggested that in a group of analyses of marine sediments, the highest correlation coefficients between element concentrations would be between those elements which occur in the same mineral. However, they also noted that fluctuations in the concentration of one major constituent could produce correlations between mineralogically unrelated elements, simply because such elements are each negatively correlated with the major constituent. Of the major phases in the present sediments, calcium carbonate varies most in abundance and is likely to be largely responsible for the positive correlations between those remaining constituents with which it is negatively correlated. The positive correlation of Sr and Mn with $CaCO_3$ would suggest that these elements are, at least in part, associated with the carbonate minerals. Such an association is well documented in the case of Sr (Thompson and Chow 1955; Turekian and Kulp 1956; Goldberg and Arrhenius 1958; Arr-

TABLE 3 - Correlation coefficients for elements in 115 sediments from the central north-eastern Irish Sea.

	$CaCO_3$	B	Ba	Cr	Mn	Sr	Ti	Zr	Fe	K	P
$CaCO_3$	1.00										
B	-0.43	1.00									
Ba	-0.35	0.69	1.00								
Cr	- 0.20	0.37	0.06	1.00							
Mn	0.52	-0.01	0.05	- 0.03	1.00						
Sr	0.82	-0.31	-0.37	-0.04	0.48	1.00					
Ti	-0.19	0.43	0.24	0.25	-0.03	-0.01	1.00				
Zr	-0.32	0.46	0.34	0.61	-0.04	-0.20	0.37	1.00			
Fe	-0.14	0.59	0.61	-0.05	0.24	- 0.13	0.46	0.20	1.00		
K	- 0.64	0.79	0.76	0.18	-0.20	-0.52	0.41	0.47	0.61	1.00	
P	0.01	0.66	0.63	0.12	0.29	-0.02	0.43	0.28	0.78	0.61	1.00

TABLE 4 - Values of the correlation coefficient for elements in central north-eastern Irish Sea sediments against mean size.

$CaCO_3$	B	Ba	Cr	Mn	Sr	Ti	Zr	Fe	K	P
0.53	-0.26	-0.31	-0.10	0.22	0.61	-0.16	-0.19	-0.03	-0.42	-0.02

henius 1963; Müller 1967), but less well established in the case of Mn although Goldberg and Arrhenius (1958) and Wangersky and Joensuu (1964) have reported its presence in the carbonate phases of some marine sediments. It has also been found in recent carbonate shells by Pilkey and Goodell (1963) and in the present study. However, the correlation between Mn and $CaCO_3$ could also, in part, be a result of the deposition of authigenic manganese dioxide phases under the high energy conditions and low detrital sedimentation rates favouring carbonate accumulation. This will be discussed in the section dealing with the distribution of manganese in the sediments.

Relationship between Composition and Grain Size

In an attempt to determine possible relationships between grain size of the sediments and their chemical composition, correlation coefficients have been calculated between their mean size and all the chemical constituents determined (Table 4). At the 0.01 significance level, element behaviour in relation to mean size can be divided into three categories. Elements such as Cr, Ti, Zr, Fe and P show no significant correlation with size, $CaCO_3$, Mn and Sr are positively correlated with size, while a negative correlation is observed in the case of B, Ba and K. These correlations can be largely explained in term of the partition of chemical constituents between the various grain size fractions of the sediments. For example, it has been noted previously that $CaCO_3$ is most abundant in the coarse grain size fractions and least abundant in the fine sediments. Similarly, the negative correlations of B, Ba and K with size can be explained in terms of their concentration in the clay fraction of the sediments (Table 2). The elements that show no positive or negative correlation with size, are likely to be fairly evenly apportioned between the various grain size fractions of the sediments, or concentrated in those of intermediate grain size, or concentrated in both the coarse and fine fractions relative to those of intermediate size. Each element will be discussed individually in the next section.

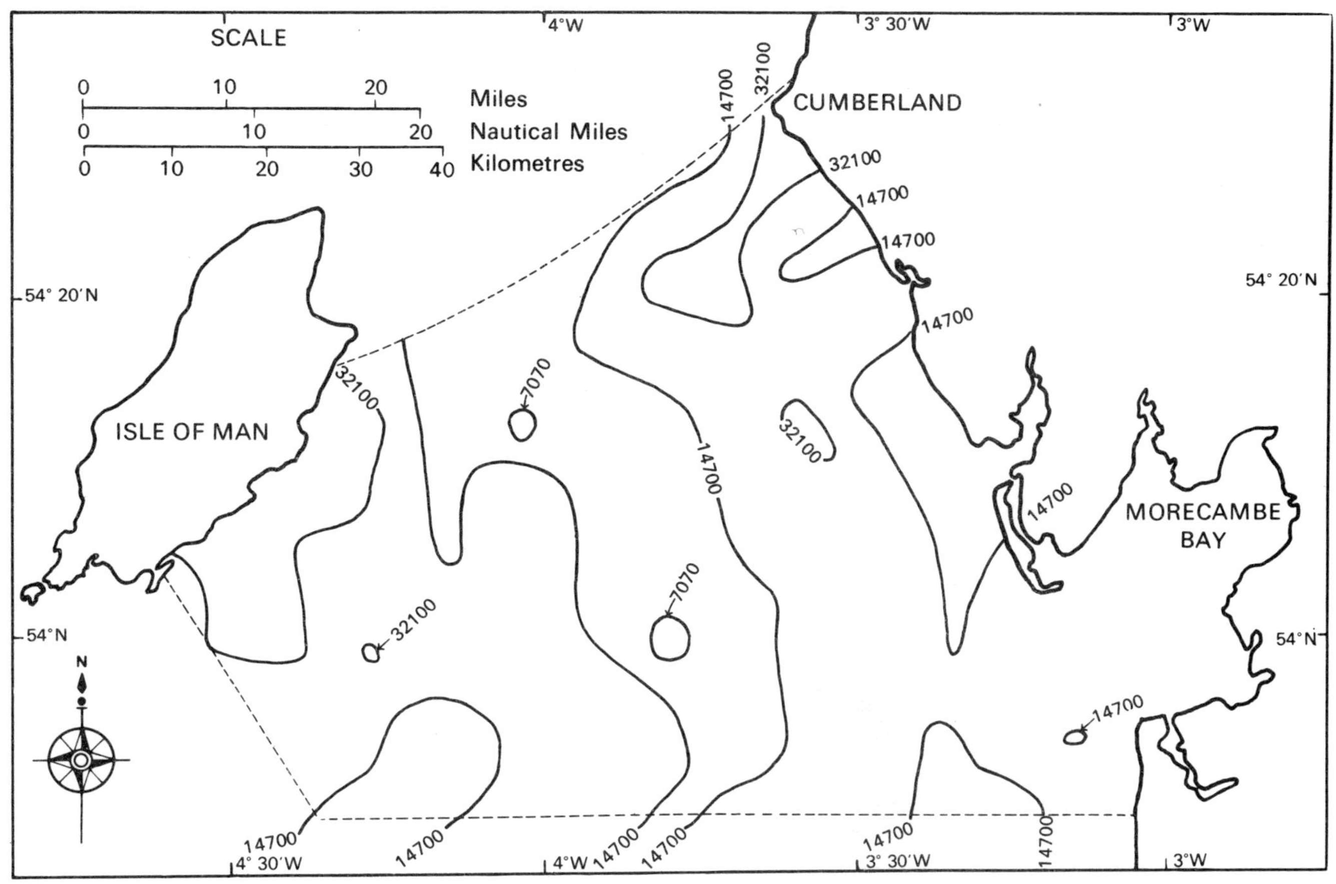

Fig. 4 - Distribution of iron (in ppm, $CaCO_3$-free, logarithmic contour intervals).

ELEMENT DISTRIBUTIONS

In this section, the distribution of each element determined in the central north-eastern Irish Sea sediments is described and briefly discussed. For the sake of simplicity, the distributions of those elements not considered to be present to any major extent in the carbonate fraction of the sediments are presented on a carbonate-free basis. All data are plotted in parts per million using a logarithmic contour interval. The latter was chosen since, with the exception of K, all of the elements determined in the sediments approach a log-normal rather than normal distribution on both a $CaCO_3$-free and total sample weight basis.

IRON

From a comparison of Figures 3 and 4 it is evident that on a carbonate-free basis, variations in the iron content of the sediments follow variations in grain size. Iron ranges in concentration between 3 and 5 per cent in the coarse deposits near the Isle of Man and decreases at first as the sediments become finer towards the east, but then increases again. It averages 0.95 per cent in the area of sand deposition in the centre of the basin, a value similar to the Fe average in sandstones of 0.98 per cent reported by Turekian and Wedephol (1961). Over much of the eastern portion of the basin it varies between 1.5 and 3 per cent, and is locally higher in sediments with a high clay content. Iron is thus concentrated in both the fine and coarse sediments, but depleted in those of intermediate grain size.

These variations can be explained in terms of the mineralogical nature of the phases concentrated in the various grain size fractions. The relatively high iron concentrations in the west of the basin are probably due mainly to the abundance of reworked glacial debris in this area. According to Pettijohn (1957), such debris is characterised by an assortment of unweathered constituents which can be derived from almost any sort of rock. It is likely therefore to contain significant concentrations of unstable iron bearing minerals which would not survive normal chemical weathering processes. Pettijohn (1957) found the average iron content of five glacial tills to be 4.09 per cent, while Angino (1966) reported an average iron content of 3.72 per cent for Antarctic glacial marine sediments closely resembling glacial tills. Furthermore, two Scottish boulder clays analysed by the Geological Survey averaged 4.38 per cent Fe (Geological Survey Wartime Pamphlet, 47, 1946 p. 34).

The abundant shell debris in the west of the area forms a contributary source of iron in the coarse sediments. Analysis of the carbonate fraction of one sample together with three different types of shell material show significant Fe concentrations in each

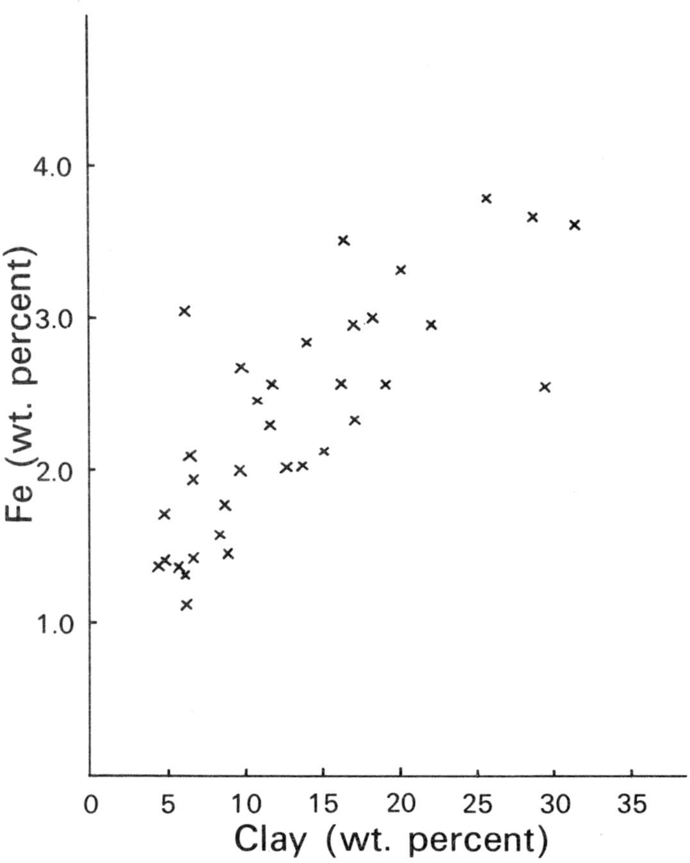

Fig. 5 - Plot of Fe against clay content.

(Table 5). These results are in general agreement with those of previous workers. Pilkey and Goodell (1963) found Fe values ranging from 5 to 1059 ppm in recent mollusc shells, while Phillips (1922) found 98 ppm Fe in an undetermined bryozoan.

	Weight per cent
-8 mesh $CaCO_3$ fraction	0.35
Lithothamnium calcareum, Areschouj	0.12
Lithothamnium glaciale, Kjellman	0.19
Glycymeris glycymeris (Linné)	0.01

TABLE 5. Iron concentrations in selected carbonate phases.

Over the remainder of the area, iron content varies with the sand/mud ratio. It is low in the sands from the centre of the basin but increases with the clay content (Fig. 5) and reaches a maximum in the black muds from the east of the basin. According to Carroll (1958), iron is associated with clays in several different ways. It can be an essential constituent of several clay minerals, a minor constituent in isomorphous substitution for other elements within clay mineral lattices, or it may occur as iron oxide coatings on the surfaces of the mineral platelets. It is an essential constituent of illite, the most abundant clay mineral in the present sediments, where it is generally present in the octahedral layer but can also occur in the tetrahedral positions. Iron is also present in chlorite (McMurchy 1934). According to Grim and Rowland (1942), illite contains an average iron content of 4.48 per cent, lower than the 6.33 per cent found in the clay fraction of the present sediments (Table 2). This difference indicates that there is more iron in the present clays than can be accomodated in the lattices of the illite group minerals. The remainder is likely to be present largely in the form of black ferrous oxide and possibly sulphide coatings on the surfaces of the mineral grains.

TITANIUM

In many respects, the distribution of titanium in the sediments (Fig. 6) follows that of iron. Similar observations have been made by Macdougall and Harriss (1969) in the fraction (smaller than 63μ) of sediments from the Canadian Arctic continental shelf, and by El Wakeel and Riley (1961) in deep-sea sediments. Like iron, titanium is moderately concentrated in the reworked glacial deposits in the western portion of the basin and decreases in concentration towards the centre, reaching a minimum in the area of medium to fine sand deposition. Like iron, it increases from the the centre of the basin eastwards, partly as a result of its enrichment in the clay fraction of the sediments (Table 2) where it may be present in the form of clay-sized Ti and Fe-bearing minerals such as ilmenite, as hydrogenous Ti precipitated from seawater or scavenged by hydrous iron oxide (Goldberg 1954), or structurally bound in the lattices of the clay minerals themselves (Hirst 1962). Macdougall and Harriss (1969) considered that the association between Fe and Ti in Canadian Arctic sediments was brought about either in the marine environment or during transport to it, and was not inherited from an association between Fe and Ti in the source rocks from which the sediments were derived. The same may be partly true in the fine fraction of the present sediments. However, in the coarser material, the association will also result from the influx of Fe and Ti-bearing detrital minerals, such as ilmenite, to the area of deposition.

According to Moore (1968), detrital titanium in Cardigan Bay sediments attained a maximum of over 2000 ppm in areas of current-swept fine grained sands. In the present sediments, it behaves in a different manner. It is lowest in the current-swept sands suggesting that titanium-rich heavy minerals such as rutile, anatase, brookite and ilmenite are at the most very scarce in these deposits. However, the greater amounts of titanium in the deposits off the Cumberland and Lancashire coasts and in the coarse sediments from adjacent to the Isle of Man may be the result of heavy mineral concentrations. In the former, it may not simply reflect the abundance of clay in the deposits since the highest titanium concentrations do not occur in the area of maximum mud accumulation. Per-

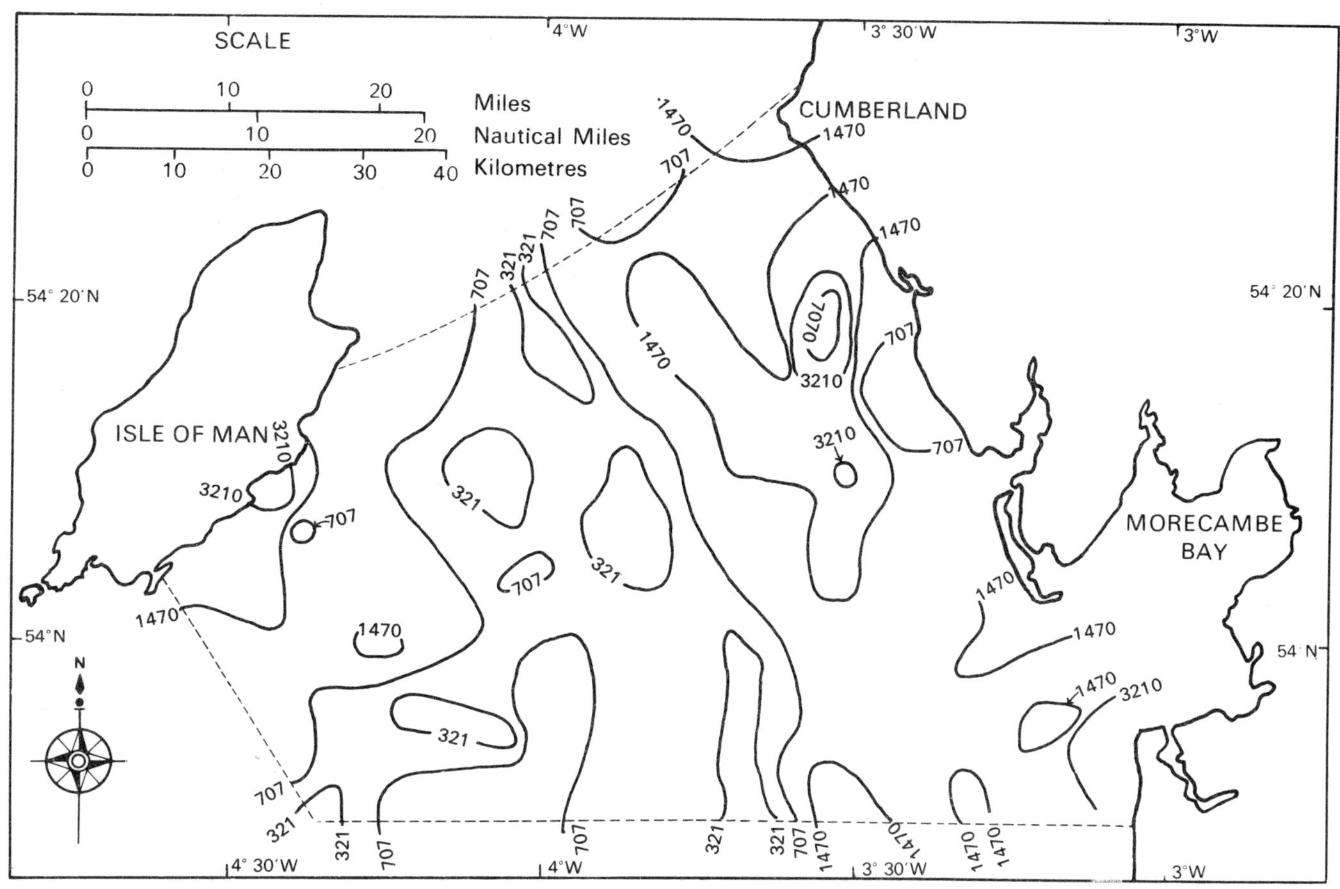

Fig. 6 - Distribution of titanium (in ppm, $CaCO_3$-free, logarithmic contour intervals).

haps local supplies of Ti-bearing heavy minerals from the English mainland are the source of the concentrations.

CHROMIUM

The distribution of chromium is rather irregular (Fig. 7). It is too scarce for detection by the analytical methods used in this work over much of the western portion of the basin, but is more concentrated in a few samples from the central zones of well sorted slightly gravelly sands and in some of the muddy sediments from the east. It is higher overall in the east than in the west at least partly as a result of its concentration in the clay fraction of the sediments (Table 2), where it may be present to some extent in the lattices of the illite group minerals (Fröhlich 1960; Hirst 1962). However, there is no direct correlation of Cr with Fe in the clays nor between Cr and the total clay content of the muddy sediments, suggesting that at least some of the chromium in the clay fraction is associated with phases other than clay minerals. Weber and Middleton (1961) consider that much of the chromium in detrital sediments is present in the form of accessory minerals such as chromite, magnetite and ilmenite. The low Fe and Ti concentrations in the centre of the basin suggest that ilmenite and magnetite are not present to any significant extent in the chromium-rich samples from this area, although chromite might be. However, all three minerals may occur in some of the deposits from the east of the basin.

ZIRCONIUM

The distribution of zirconium is irregular throughout the area (Fig. 8) and probably reflects the abundance of zircon in the sediments. However, under certain circumstances zirconium is slightly soluble and can be incorporated into developing clay minerals (Goldschmidt 1954; Dengenhardt 1957; Rankama and Sahama 1950), although the low zirconium content of the clays examined in this work (Table 2) suggests that this mechanism is of minor importance in the present sediments. Zirconium was not detected over much of the western portion of the basin, although there are local concentrations, but it is more abundant in the centre and is highest in the muddy sediments from the east. Since it is depleted in the clay fraction of the sediments relative to its concentration in the other grain-size classes (Table 2) its enrichment in the muddy deposits cannot be a function of the amount of clay they contain. The lower concentrations of Zr in the muds and sandy muds compared with the muddy sands (Table 7) suggest that it exists in these sediments largely in silt and fine sand-sized particles. The area of greatest zirconium content is in the south-east of the basin where low-energy muddy sands are accumulating. Here values of over 400 ppm Zr are not uncommon, and they increase locally to over 2000 ppm adjacent to the Lancashire coast. Such an association does not support the assumption of Moore (1968) that Zr concentrations in offshore sediments indicate high energy sedimentation.

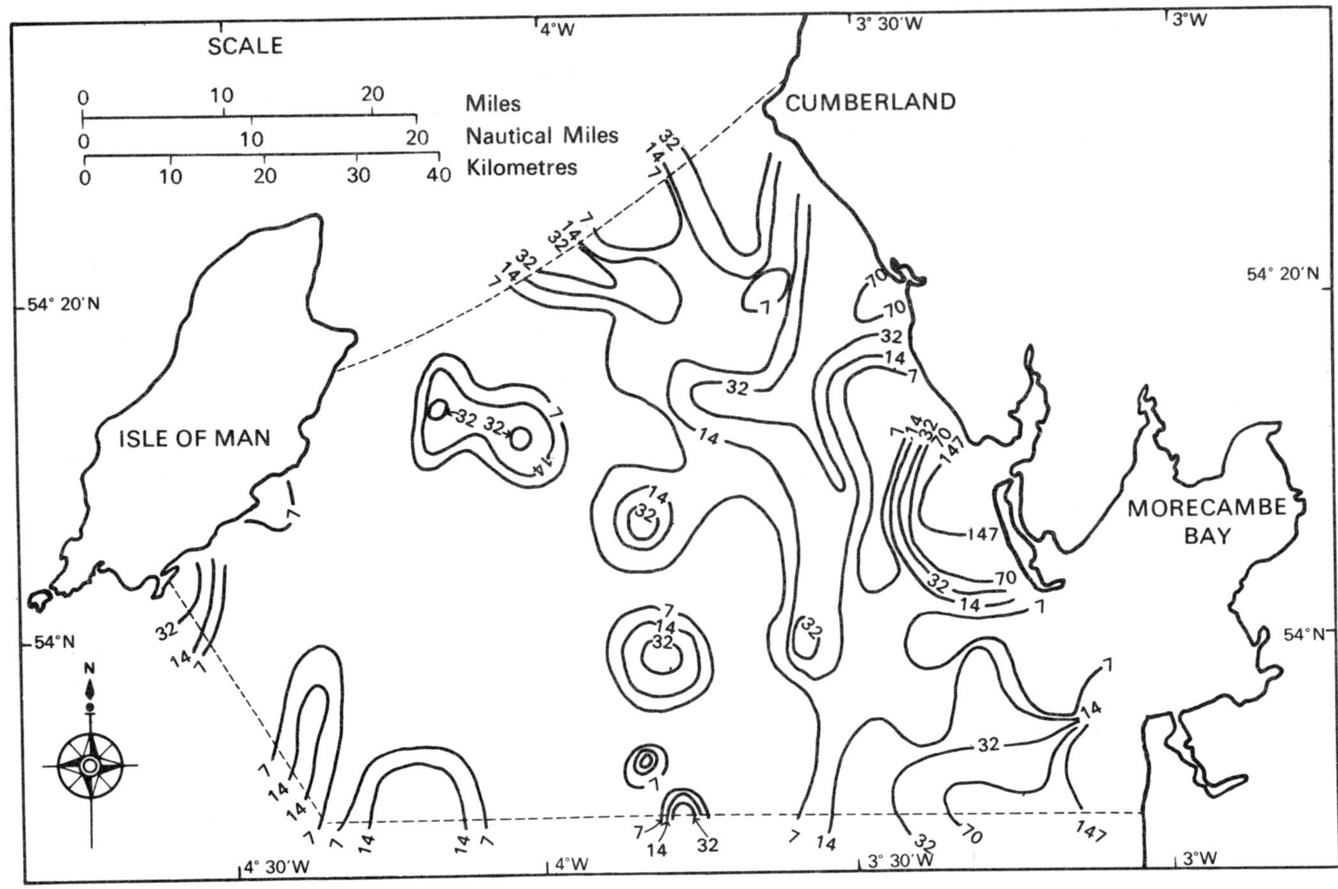

Fig. 7 - Distribution of chromium (in ppm, $CaCO_3$-free, logarithmic contour intervals).

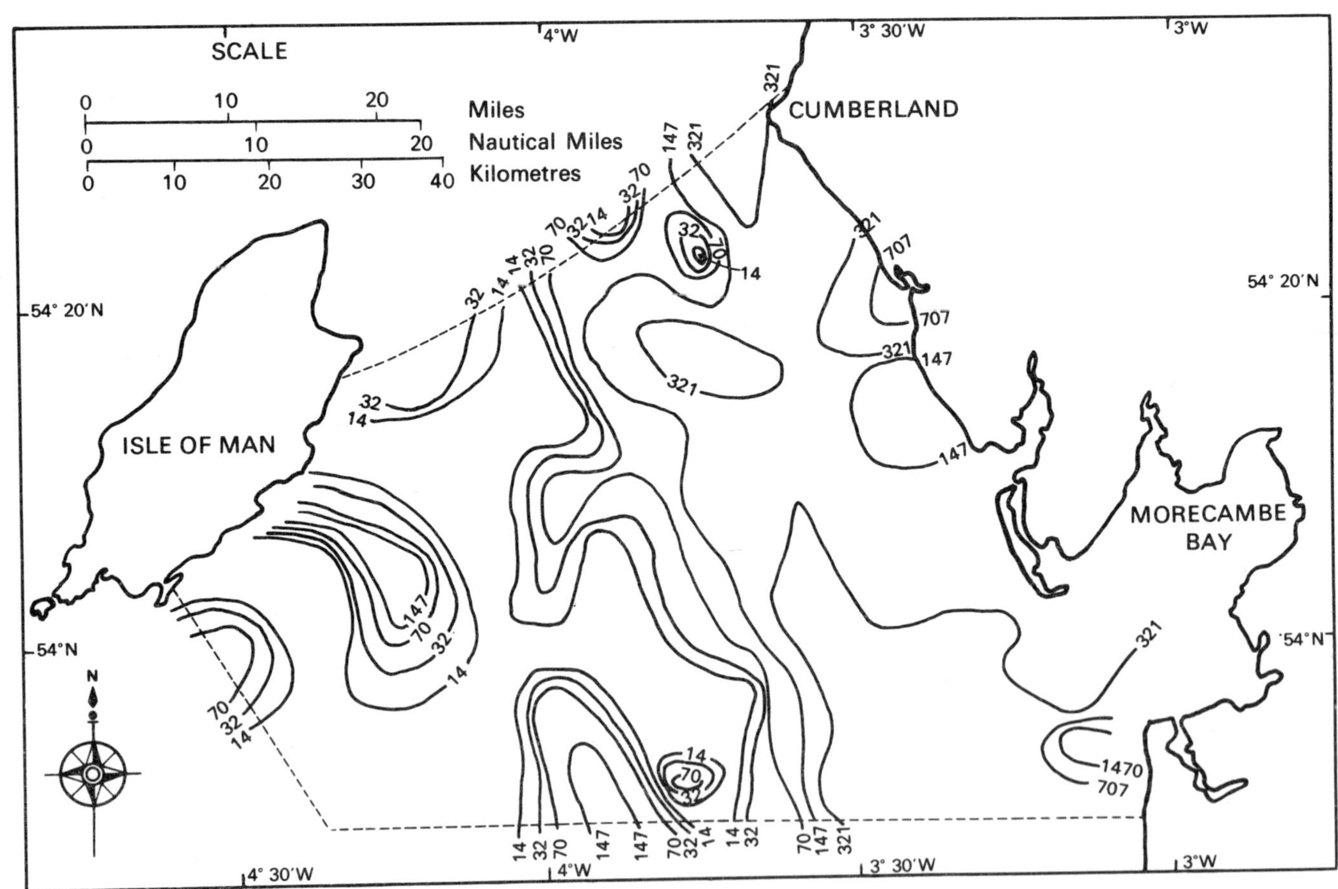

Fig. 8 - Distribution of zirconium (in ppm, $CaCO_3$-free, logarithmic contour intervals).

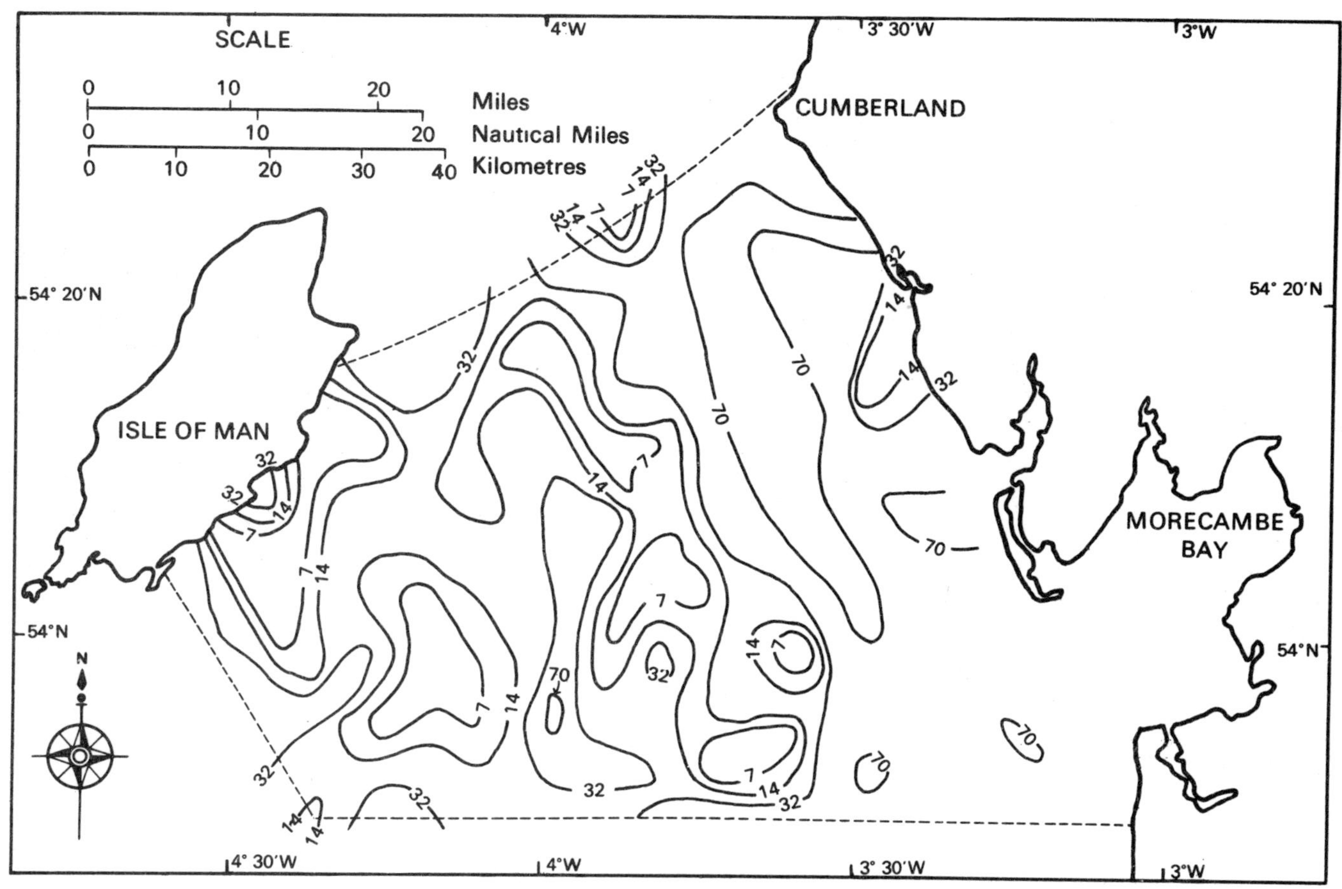

Fig. 9 - Distribution of boron (in ppm, $CaCO_3$-free logarithmic contour intervals).

BORON

The sedimentary geochemistry of boron has received considerable attention in recent years because of the possible use of this element as a palaeosalinity indicator. Its association with clays is well known (Degens 1965) and as would be expected it is considerably more abundant in the clay fraction of the present sediments than in the other grain-size classes (Table 2). Over much of the western part of the survey area it is present in the sediments in concentrations of less than 25 ppm although it is locally abundant close to the Isle of Man (Fig. 9). It is also concentrated locally in the gravelly and slightly gravelly sands from near the centre of the basin. Values here range between 10 and 60 ppm in comparison with a boron average in sandstones of 35 ppm (Turekian and Wedephol 1961), and reach 70 ppm in one outcrop of muddy sandy gravel. These high values suggest a concentration of boron bearing minerals in the deposits. Since there is very little clay in these sediments, the boron must be present in another mineral phase. Hirst (1962) and Moore (1968) have found boron in detrital tourmaline and/or muscovite in shallow marine sands, and it is possibly present in one or both of these phases in the sand fraction of the present sediments.

There is a general increase in the abundance of boron with increase in the clay content of the muddy sediments from the eastern portion of the basin (Fig. 10). However, there are large fluctuations in B concentrations not associated with variations in clay

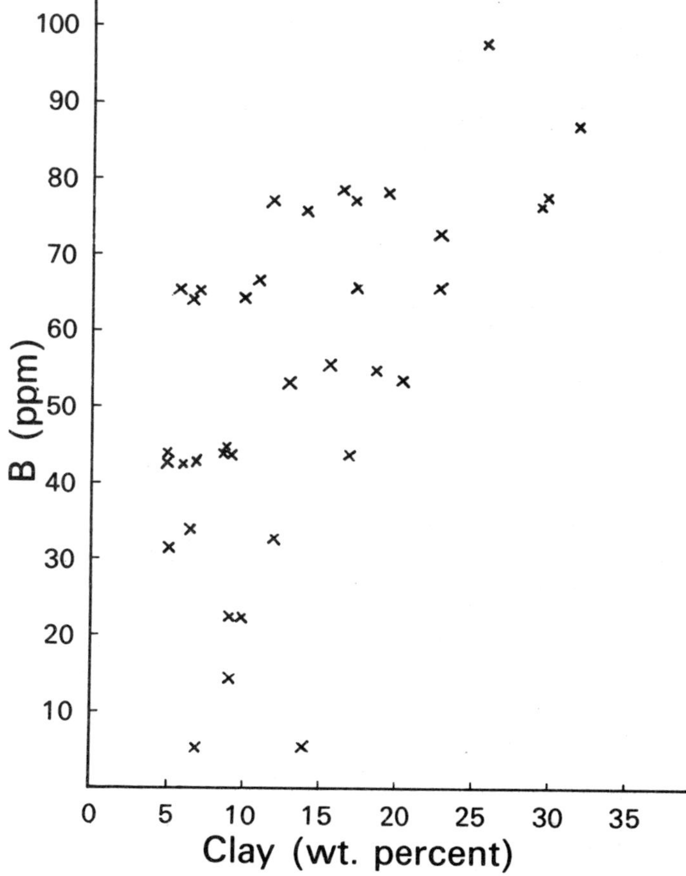

Fig. 10 - Plot of B against clay content.

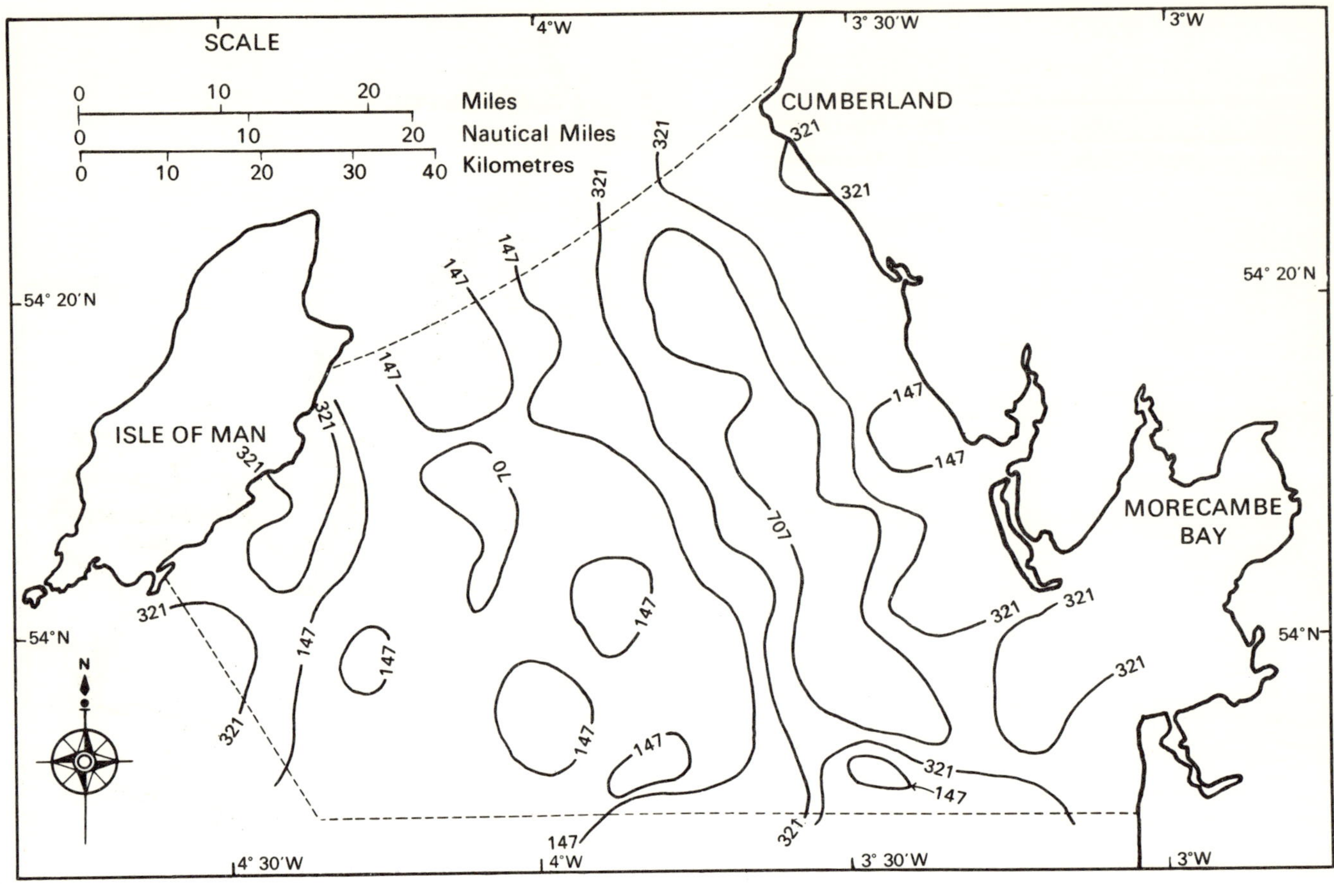

Fig. 11 - Distribution of barium (in ppm, $CaCO_3$-free, logarithmic contour intervals).

content, suggesting that phases other than clays may contribute boron to these sediments. Nevertheless, much of the boron is probably associated with the clay minerals, either substituting in their lattices or more loosely held by adsorption. Boron in the clay mineral lattices will have largely entered the basin of deposition fixed in this position (Hirst 1962), while the adsorbed boron will have been extracted from seawater during the present sedimentation cycle. Harder (1961) suggested that illite may take in boron more readily than other clay minerals, but Levinson and Ludwick (1966) noted that the finest clay particles adsorb the most boron irrespective of their mineralogical nature.

BARIUM

The lowest barium concentrations in the sediments are found in the zones of gravelly and slightly gravelly sands in the north-west of the area (Fig. 11). Here values of less than 70 ppm occur although over 300 ppm Ba is found in some of the coarser gravelly sands adjacent to the Isle of Man. According to Moore (1968), barium in Cardigan Bay sands is related to the distribution of phyllosilicates, feldspars and accessory minerals, while Weber and Middleton (1961) found it to be related to feldspars, carbonates and clays in the sediments they examined. Feldspars could be partly responsible for the slightly higher amounts of Ba in the sediments adjacent to the Isle of Man, since the detrital fraction of these sediments consists largely of reworked glacial debris which probably contains a greater proportion of feldspars than the better sorted sandy deposits to the east. The abundant shell debris near the Isle of Man could also be a contributary source of Ba in these deposits. Pilkey and Goodell (1963) reported concentrations of up to 48 ppm Ba in recent mollusc shells from marine environments.

In the muddy sediments from the east of the basin, barium concentrations are considerably higher than in the west. Values of over 700 ppm Ba are common over much of this region and locally increase to over 1000 ppm in the area of maximum mud deposition. These high values will be due principally to the concentration of Ba in the clay fraction of the sediments (Table 2) where it may substitute for K in the interlayer position of the illite group clay minerals. However, according to Rankama and Sahama (1950), some Ba may also be adsorbed onto clays from seawater and enter into exchange positions.

POTASSIUM

Potassium is concentrated in the clay fraction of the sediments (Table 2) where it is probably largely present in the illite group minerals. It is most abundant in the muddy sediments from the east of the basin (Fig. 12) and is correlated with their clay content (Fig. 13). The generally low K values in sediments from the west and centre of the basin (Fig. 12)

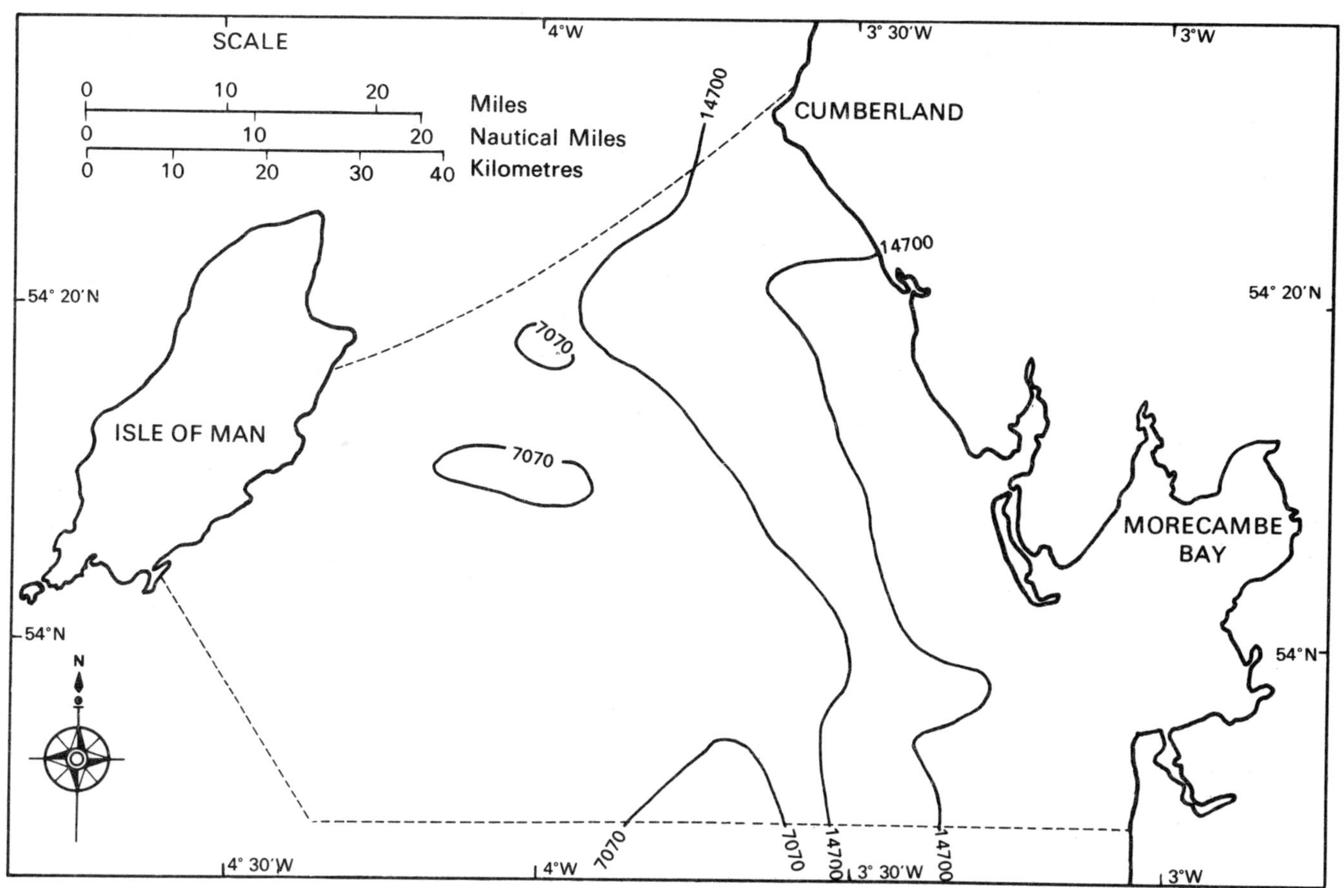

Fig. 12 - Distribution of potassium (in ppm, $CaCO_3$-free, logarithmic contour intervals).

can be related to the virtual absence of clay in these deposits. Potassium concentrations of between 1.0 and 1.4 per cent are not uncommon in the reworked glacial deposits adjacent to the Isle of Man, but are less than 1 per cent over the gravelly, slightly gravelly and well sorted sands to the east. According to Moore (1963), in Buzzards Bay, Massachusetts, potassium not related to clays, can be related to the distribution of feldspars. A similar relationship may occur in the present sediments since feldspars are likely to be more abundant in the glacial deposits adjacent to the Isle of Man than in the sandy deposits to the east.

PHOSPHORUS

Phosphorus is most abundant on a carbonate-free basis in the coarse glacial and calcareous deposits from the west of the basin (Fig. 14). However, its actual concentration in these sediments only ranges between about 350 and 500 ppm, some of which may be located in the carbonate skeletal remains (Clarke and Wheeler 1917; El Wakeel and Riley 1961). Phosphorus decreases in concentration towards the centre of the basin and reaches a minimum of between 200 and 300 ppm in the medium to fine sands. Farther east it increases again as the proportion of mud in the sediments increases. This is partly a result of its concentration in the clay fraction of the sediments (Table 2) where it may be present in a variety of phases including the clay minerals, finely divided authigenic and detrital calcium phosphate, faecal pellets and other organic remains.

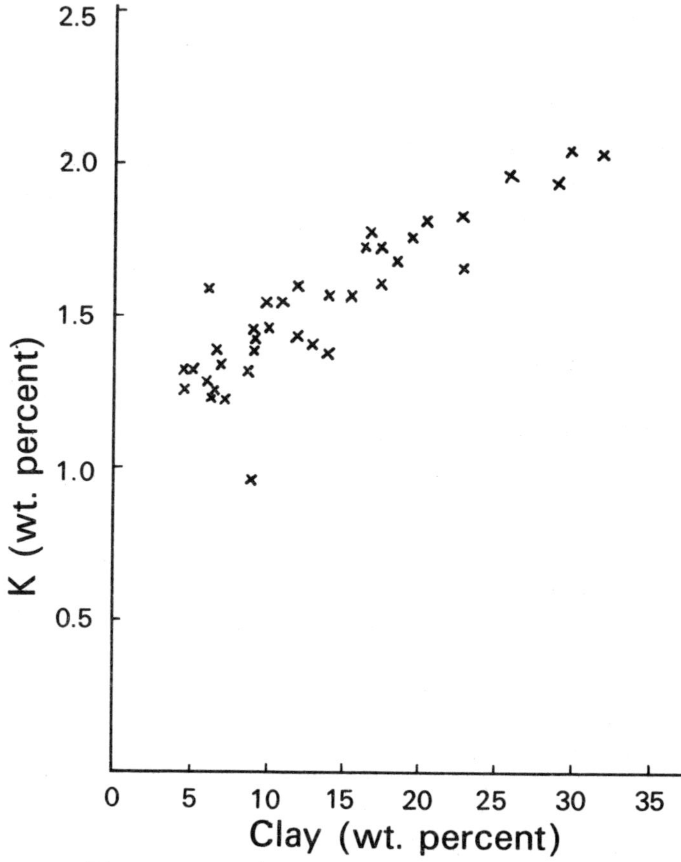

Fig. 13 - Plot of K against clay content.

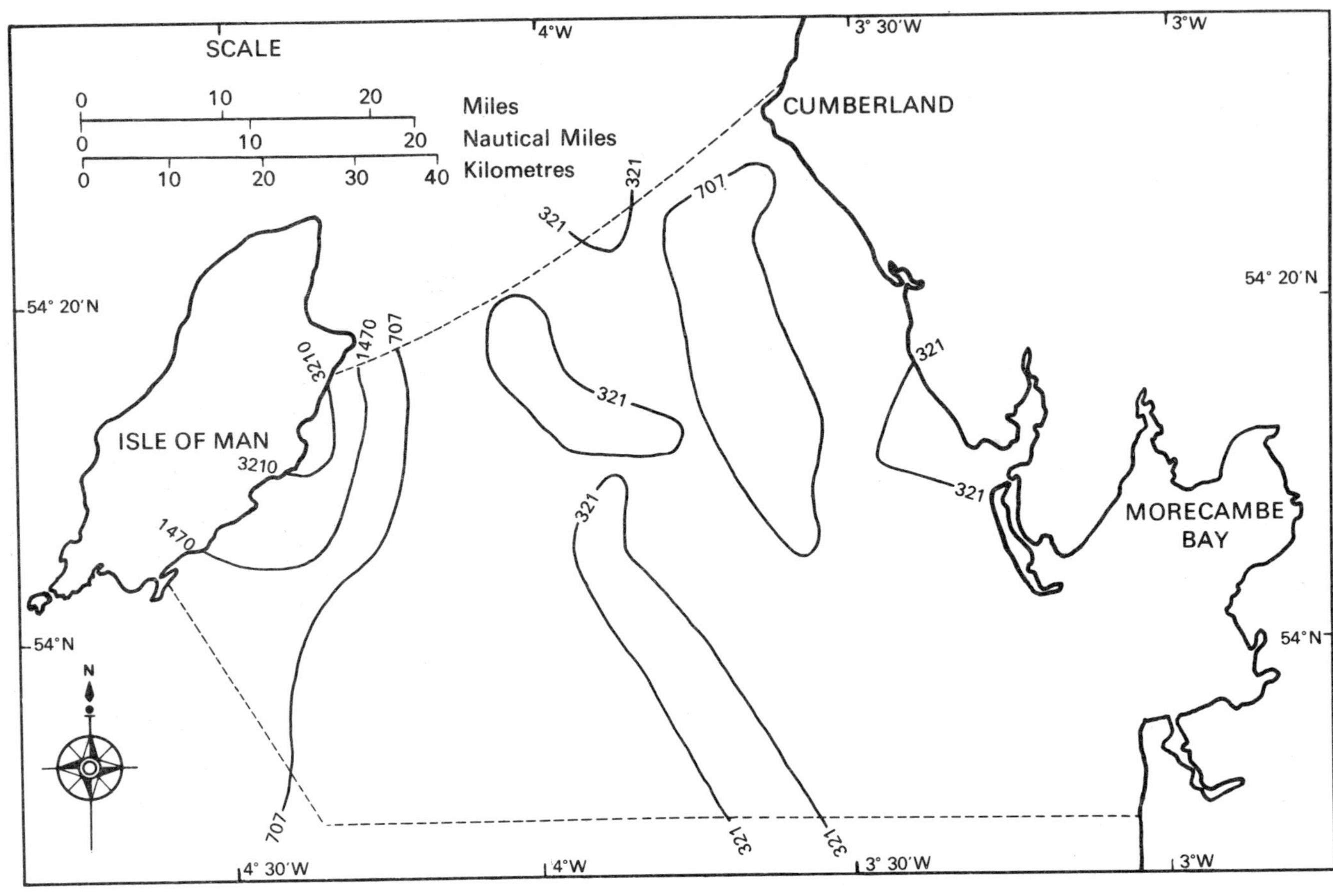

Fig. 14 - Distribution of phosphorus (in ppm, $CaCO_3$-free, logarithmic contour intervals.)

STRONTIUM

Figures 15 and 3 show that the distribution of strontium in the area largely follows that of calcium carbonate. This relationship is further illustrated in Table 3 by the correlation coefficient between the two and suggests that strontium is principally present in the sediments associated with carbonate phases Several workers have noted an enrichment of Sr in carbonate minerals and that it is more concentrated in aragonite than in calcite (Bowen 1956; Thompson and Chow 1955; Graf 1960). According to Degens (1965), this difference is largely structurally controlled since strontianite ($SrCO_3$) is isostructural with aragonite but not with calcite. Turekian and Kulp (1956) found that Sr in calcite rarely exceeded 610 ppm, but Hirst (1962) found 1650 ppm in aragonite shell debris from the Gulf of Paria, and Pilkey and Goodell (1963) recorded values of up to 3318 ppm Sr in recent mollusc shells composed of aragonite. The occurrence of Sr in concentrations of up to 2000 ppm in some of the calcareous shelly deposits from the west of the present area suggests that these too are predominantly aragonitic in composition.

As the calcium carbonate content of the sediments decreases from west to east across the basin, so does the strontium content. However, certain samples, especially from the east of the basin, contain more Sr than the average (Fig. 15) but are not rich in $CaCO_3$. This indicates that carbonates are not the only control on the abundance of strontium in the sediments. Some strontium will enter the basin of deposition structurally bound in clay minerals and some will be adsorbed onto the clays from seawater. However, a comparison of the average Sr content of the bulk samples with that of the clays (Table 2) shows these amounts to be of minor importance. The average Sr content of the latter is 111 ppm, well within the range of previously reported average Sr values for the clay fraction of Atlantic deep-sea sediments (50 - 250 ppm; Turekian 1964), indicating that little or no fractionation of Sr occurs between the present clays and those deposited in deep-sea environments. The removal of strontium from seawater is probably largely brought about by organic mechanisms, and its extraction from the oceans by cation exchange processes on clays is likely to be of minor importance irrespective of the nature of the depositional environment of the sediments.

MANGANESE

Manganese in the sediments is positively correlated with calcium carbonate and, like strontium, its distribution (Fig. 16) may be influenced partly by the distribution of the carbonate phases. Pilkey and Goodell (1963) found concentrations of up to 438 ppm Mn in recent mollusc shells, and other occurrences of Mn in marine carbonate organisms were reviewed by Vinogradov (1953) and Graf (1960). Analyses of carbonate phases from the west of the present area have found Mn in concentrations of up

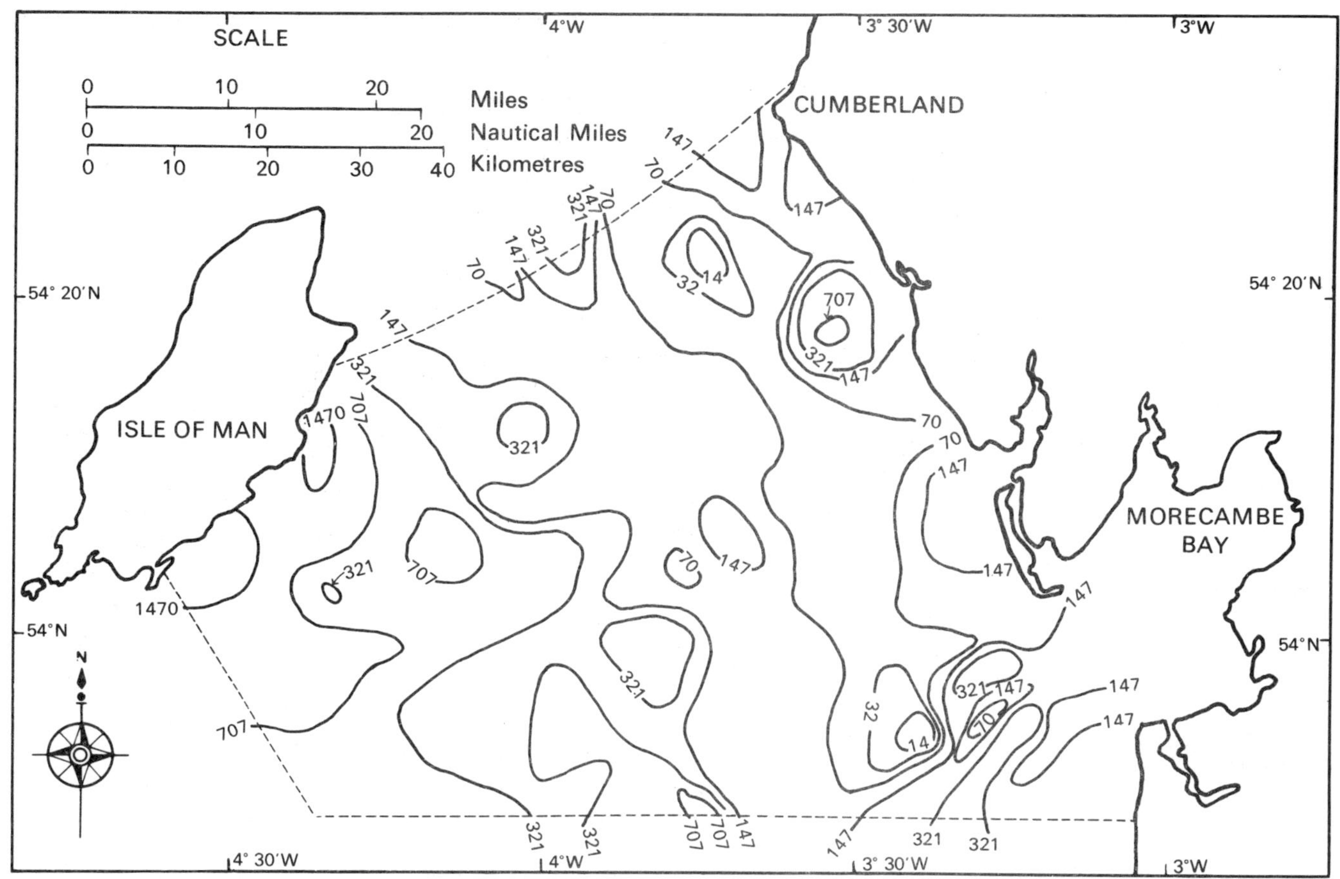

Fig. 15 - Distribution of strontium (in ppm, logarithmic contour intervals).

to 1500 ppm (Table 6), indicating that a large part of the total manganese in these sediments is associated with the carbonate fraction. Manganese may substitute as Mn^{2+} for Ca^{2+} in the lattices of the carbonate minerals. Wildeman (1970) reported that the results from the electron paramagnetic resonance (E.P.R.) spectra of manganese in natural and synthetic carbonate minerals show that manganese exists in these phases as Mn^{2+} dispersed throughout the carbonate lattice. However, according to Graf (1960), carbonate skeletal material may contain significant proportions of structural protein or its degradation products, and minor elements associated with carbonates could partly be bound up in these phases.

	Weight per cent
-8 mesh $CaCO_3$ fraction	0.1000
Lithothamnium calcareum, Areschouj	0.0900
Lithothamnium glaciale, Kjellman	0.1500
Glycymeris glycymeris (Linné)	0.0026

TABLE 6. Manganese concentrations in selected carbonate phases.

Manganese not present in the carbonate fraction of sediments from the western portion of the basin is likely to be located mainly in detrital minerals and rock fragments, although some may occur in the form of authigenic ferromanganese oxide phases. Brown coatings were observed on the surfaces of several shells from the west of the basin, and under the conditions of high current velocities and negligible detrital sedimentation in this area (Cronan 1969), the precipitation of ferromanganese oxide phases as coatings on mineral grains and exposed surfaces is by no means unlikely. Such deposits have been found in a variety of continental shelf environments (Manheim 1965) and abound on the deep-sea floor.

As the grain size of the sediments decreases in an easterly direction, their manganese content also decreases and reaches a minimum in the area of sand deposition. Here, manganese is present in concentrations of between 200 and 300 ppm. With further easterly decrease in grain size, manganese increases and reaches a secondary maximum in the mud deposits in the north-east of the basin. It is largely concentrated in the clay fraction of these sediments (Table 2) and may be present in the clay mineral lattices and/or adsorbed onto the surfaces of the clay particles. The Eh of the environment in this area is probably too low for manganese to be oxidised to the trivalent or quadrivalent state and precipitated as an insoluble oxide.

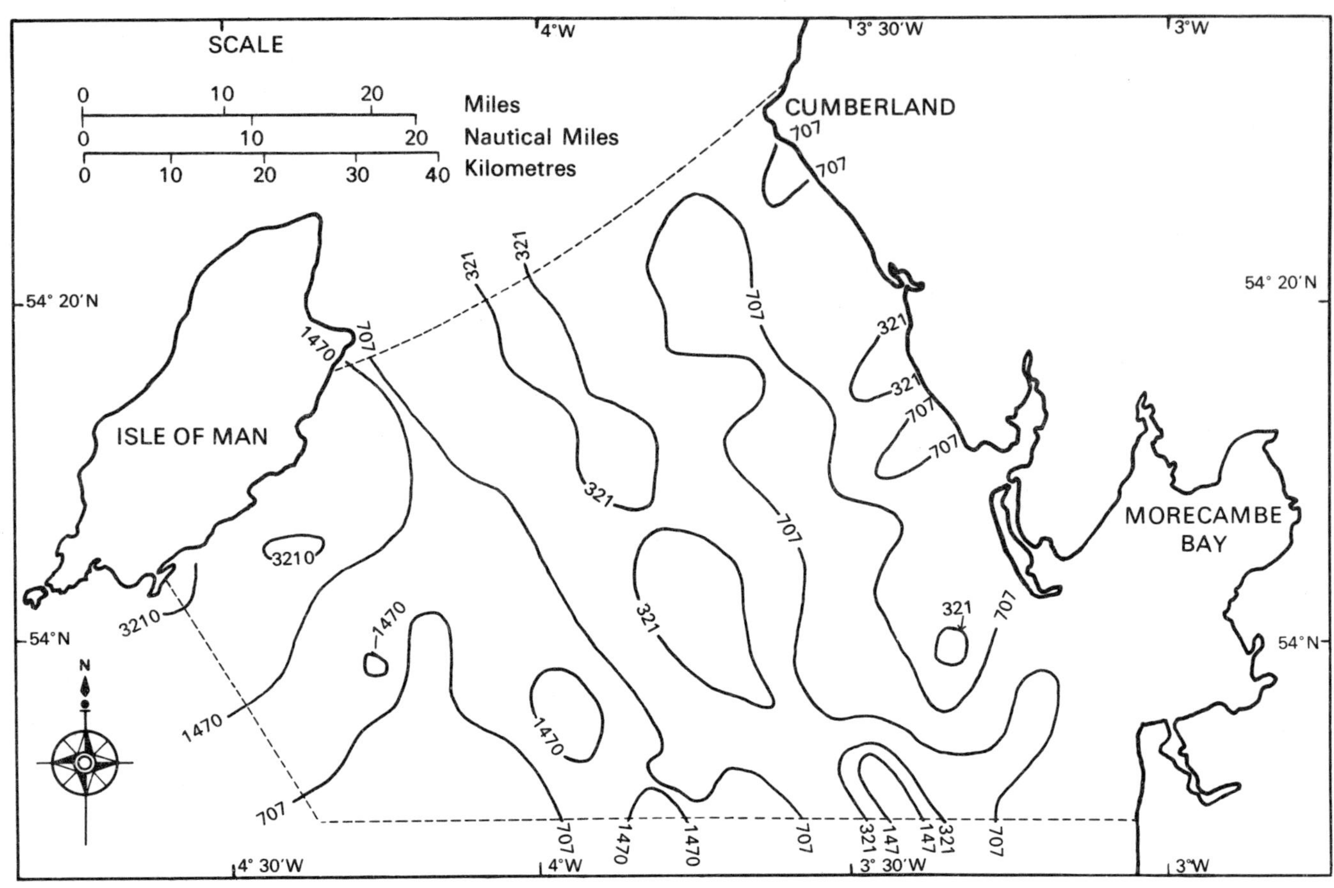

Fig. 16 - Distribution of manganese (in ppm, $CaCO_3$-free, logarithmic contour intervals).

GALLIUM

Gallium is present in all the bulk sediments analysed in concentrations of less than 10 ppm, but was found in concentrations of up to 20 ppm in the clay fractions. According to Hirst (1962), the sedimentary geochemistry of gallium is largely influenced by the similarity of Ga^{3+}, with ionic radius 0.62Å, to Al^{3+} with ionic radius 0.51Å. He found the Ga/Al ratio in sediments from the Gulf of Paria to be remarkably constant and considered that this was due to Ga entering the basin of deposition structurally combined within the lattices of clay minerals. Much of the gallium was found to be replacing aluminium in kaolinite, with lesser amounts in illite and montmorillonite. The average Ga content of the present clays is 8 ppm, lower than the 22 ppm occuring in Gulf of Paria clays, This difference may be a reflection of the predominantly illitic character of the north-eastern Irish Sea clays and their low content of kaolinite.

COBALT, COPPER, NICKEL, VANADIUM AND ZINC

In gèneral, these elements are most abundant in the muddy sediments from the east of the basin and are lowest in the sands from the centre. On average, all are present in the Irish Sea sediments in somewhat lower concentrations than in sediments from the other near-shore environments listed in Table 2, and very much lower than in those from the deep sea. A proportion of all of the elements probably entered the basin of deposition bound structurally in the lattices of the detrital clay minerals (Hirst 1962), but some may have also been incorporated in the clay fraction by adsorption from seawater (Chester 1965). A small amount may also be present in the coarser grained detrital minerals, especially in the reworked glacial debris in the west of the area. Most of the elements are more abundant in this debris than in the sands and some approach the concentration in which they occur in the muddy sediments to the east. Moore (1968) noted that Co, Cu, Ni and V were largely associated with phyllosilicates and accessory minerals in Cardigan Bay sediments, and similar conclusions have been drawn by Weber and Middleton (1961) in sediments they examined.

CONCLUSIONS

This account shows that the concentrations of the elements determined in sediments from the central north-eastern Irish Sea vary largely with sediment type. This conclusion is further exemplified by a comparison of the average composition of the different sediments present in the basin classified according to the scheme of Folk (1954) (Tables 7 and 8). In Table 7 the average composition of each of the major sediment types in the basin is presented on a total weight basis. However, since the dilutive effect of calcium carbonate profoundly effects the abundances of most of the elements in the sediments, the averages are

TABLE 7 - Average composition of different sediment types in the central north-eastern Irish Sea.

	$CaCO_3$ (%)	B (ppm)	Ba (ppm)	Cr (ppm)	Mn (ppm)	Sr (ppm)	Ti (ppm)	Zr (ppm)	Fe (%)	K (%)	P (ppm)
Sandy Gravel	48.57	8	95	5	979	866	520	35	1.50	0.52	457
Gravelly Sand	31.19	15	93	12	731	531	468	30	1.10	0.56	369
Slightly Gravelly Sand	11.52	22	111	12	475	145	412	82	1.09	0.73	315
Sand	7.78	21	189	28	340	150	590	75	0.88	0.91	336
Muddy Sand	9.76	40	375	17	557	133	1342	256	1.62	1.41	452
Sandy Mud	9.05	68	575	18	675	69	1790	231	3.00	1.68	691

TABLE 8 - Average composition of different sediment types in the central north-eastern Irish Sea recalculated on a calcium-carbonate-free basis.

	B (ppm)	Ba (ppm)	Cr (ppm)	Mn (ppm)	Sr (ppm)	Ti (ppm)	Zr (ppm)	Fe (%)	K (%)	P (ppm)
Sandy Gravel	15	184	10	1899	1680	1008	68	2.91	1.01	886
Gravelly Sand	22	134	17	1059	769	678	43	1.59	0.82	535
Slightly Gravelly Sand	24	125	13	536	163	465	92	1.23	0.83	355
Sand	23	204	30	367	162	637	81	0.95	0.98	362
Muddy Sand	44	412	19	612	146	1476	281	1.78	1.56	497
Sandy Mud	74	626	20	735	75	1951	251	3.27	1.83	753

also presented on a recalculated carbonate-free basis for comparison (Table 8). The latter procedure artificially inflates the abundances of elements such as Sr which are present to a large extent in the carbonate minerals, but has the advantage that it enables comparisons to be drawn between the abundances of elements concentrated in the non-carbonate fractions of the sediments.

The non-carbonate fraction of the sediments is considered to consist mainly of material of glacial origin which has been reworked and selectively transported from west to east across the basin by tidal current action. This selective transportation has effected a grain size separation of the parent glacial material and thus the compositional variations in the non-carbonate fraction of the sediments are likely to reflect compositional variations between the various size fractions of the original tills from which the sediments were derived, with some modifications resulting from selective breakdown of unstable minerals during transport and uptake of elements from seawater.

The non-carbonate fraction of the sandy gravels in the west of the basin is considered to be a lag deposit from which most of the fine constituents have been removed. These sediments therefore probably contain a higher proportion of chemically unstable minerals than do the transported deposits to the east, since unstable minerals would be selectively broken down during submarine transport processes. This might, in part, account for their slight enrichment on a carbonate-free basis in elements such as Fe, Ti and K, relative to the transported sands, since these elements are likely to be present in significant concentrations in phases such as rock fragments, feldspars, mafics and some accessory minerals which would probably not survive significant submarine transportation. Apart from Ca, Mn and Sr are the only elements notably abundant in the sandy gravels on a total sample weight basis relative to their concentrations in the other sediments, probably because of their presence in the lattices of the carbonate minerals. However, the abundance of Mn might also be due to the precipitation of authigenic ferromanganese oxide minerals as coatings on shells and other exposed surfaces. Certain elements such as B, Cr and Zr on a carbonate-free basis are more concentrated in some of the gravelly sands and sands from the centre of the basin than in the sandy gravels to the west. This would suggest that these elements have been derived largely from the sand fraction of the parent glacial tills and are present in accessory minerals such as zircon, tourmaline and possibly chromite. However, the generally low abundance of these elements throughout the area indicates that such minerals can only be present in minor concentrations. Towards the east of the basin the increasing proportion of clay in the sediments markedly influences their composition. Elements such as B, Ba, Cr, Fe, K, Mn, P and Ti, which are concentrated in the clay fraction, become more abundant as the amount of this fraction increases. With the exception of Cr, each of these elements is more concentrated in the muddy sands than in the sands, and still more concentrated in the sandy muds. A proportion of each of them will be structurally combined in the lattices of the clay minerals, especially since the clays are thought to be largely glacial in origin and thus would not have been subjected to extensive chemical leaching in their source areas or during transport. However, the ion exchange capacity of clay minerals is well known (Carroll 1959) and a fraction of the elements present is likely to have been extracted from seawater by adsorptive processes. For example, potassium is effectively removed from dilute solutions at moderate pH by illite (Degens 1965), and Levinson and Ludwick (1966) considered that boron is adsorbed onto clay minerals when they enter the marine environment. The only elements determined in the present study which are not concentrated in the clay fraction are strontium and zirconium. The former is largely extracted from seawater by organisms secreting calcium carbonate, and since the remains of such organisms are rare in the clays their low Sr content is explained. A significant proportion of the Sr that does occur in the clay fraction of the sediments is probably substituting for Ca^{2+} in calcite, while the remainder is likely to be partitioned between lattice and ion exchange sites in the clay minerals. The concentration of zirconium in the muddy sands, and to a lesser extent in the sandy muds is probably a reflection of the generally small size of zircon crystals in primary rocks. Feniak (1944) gave the average size of zircons in intermediate and acid igneous rocks as 0.066 x 0.046 mm and 0,063 x 0.042 mm respectively.

The simple grain size fractionation and redistribution of glacial tills by tidal current action cannot account for some anomalies in the composition of sediments from the eastern margin of the basin. Elements such as Ti, Zr and Cr which are often present in sediments in significant concentrations in resistant heavy and accessory minerals, are considerably concentrated in some deposits near the English coast relative to their abundance in the eastern portion of the area as a whole. It is difficult to explain such concentrations in terms of the derivation of the minerals containing these elements from the west of the basin and their transport to the east by tidal currents. It is perhaps more likely that they have been supplied from the English mainland by coastal erosion and/or river discharge.

In conclusion therefore, the geochemistry of sediments in the central north-eastern Irish Sea is thought to be a reflection of several factors. These include (a) the mineralogy and chemistry of the various grain size fractions of the glacial tills from which the bulk of the sediments have been derived, (b) the selective breakdown of chemically unstable minerals during submarine transport, (c) the formation of authigenic phases within the basin of deposition, (d) the adsorption of elements from seawater by clay minerals and (e) the possible local supply of materials from the English mainland by coastal erosion or river discharge.

ACKNOWLEDGEMENTS

The work presented in this paper was begun while the writer was a member of the Institute and completed in the Department of Geology, University of Ottawa. Gratitude is due to several of the Institute's technical staff including Mr. R.J. Merriman, Mr. J. Dangerfield and Mr. M.J. Cope of the Petrographical Department for, respectively, the X-ray diffraction analyses, assistance in sample preparation and general assistance through the early stages of the study. Thanks are also due to Dr. S.H.U. Bowie and Mr. W.H. Bennett of the Geochemical Division through whose courtesy the chemical analyses were done, Mr. J.E. Wright and Dr. P.A. Sabine of Continental Shelf Unit I and the Petrographical Department respectively for providing valuable advice during all stages of the study, Dr. A.W. Woodland for supplying the samples and Dr. K.C. Dunham for permission to undertake the project.

REFERENCES

ALLEN, V.T. and JOHNS, W.D. 1960 Clays and clay minerals of New England and eastern Canada. *Geol. Soc. America Bull.*, **71**, 75-86.

ANGINO, E.E. 1966 Geochemistry of Antarctic pelagic sediments. *Geochim. et Cosmochim. Acta*, **30**, 939-961.

ARRHENIUS, G. 1963 Pelagic sediments. In: *The Sea: Ideas and Observations on Progress in the Study of the Seas.* M.N. Hill, editor, Interscience, New York 3, 655-727.

BERRY, R.W. and JOHNS, W.D. 1966 Mineralogy of the clay-sized fractions of some North Atlantic - Arctic ocean bottom sediments. *Geol. Soc. America Bull.*, **77**, 183-196.

BISCAYE, P.E. 1965 Mineralogy and sedimentation of recent deep-sea clay in the Atlantic Ocean and adjacent seas and oceans. *Geol. Soc. America Bull.*, 76, 803-832.

BOWEN, H.J.M. 1956 Strontium and barium in sea water and marine organisms. *Marine Biol. Assoc. [United Kingdom] Jour.*, **35**, 451-460.

CARROLL, D. 1958 Role of clay minerals in the transportation of iron. *Geochim. et Cosmochim. Acta*, **14**, 1-27.

CARROLL, D. 1959 Ion exchange in clays and other minerals. *Geol. Soc. Americal Bull.*, **70**, 749-780.

CHAVE, K.E. and MACKENZIE, F.T. 1961 A statistical technique applied to the geochemistry of pelagic muds. *J. Geol.*, **69**, 572-582.

CHESTER, R. 1965 Adsorption of zinc and cobalt on illite in sea-water. *Nature*, **206**, 884-886.

CLARKE, F.W. and WHEELER, W.C. 1917 The inorganic constituents of marine invertebrates *U.S. Geol. Survey Prof. Paper* 102.

CRONAN, D.S. 1969 Recent sedimentation in the central north-eastern Irish Sea. *Rep. No. 69/8, Inst. Geol. Sci. 10 pp.*

CRONAN, D.S. 1969a Average abundances of Mn, Fe, Ni, Co, Cu, Pb, Mo, V, Cr, Ti and P in Pacific pelagic clays. *Geochim. et Cosmochim. Acta*, **33**, 1562-1565.

DEGENS, E.T. 1965 *Geochemistry of Sediments* Prentice-Hall, New Jersey, 342 pp.

DENGENHARDT, H. 1957 Untersuchungen zur geochemischen Verteilung des Zirkoniums in der Lithosphäre. *Geochim. et Cosmochim. Acta*, **11**, 279-309.

EL WAKEEL, S. and RILEY, J.P. 1961 Chemical and mineralogical studies of deep-sea sediments. *Geochim. et Cosmochim Acta*, **25**, 110-146.

FENIAK, M.W. 1944 Grain sizes and shapes of various minerals in igneous rocks, *Amer. Mineral.*, **29**, 415-421.

FOLK, R.L. 1954 The distinction between grain size and mineral composition in sedimentary rock nomenclature. *J. Geol.*, 62, 344-359.

FRYE, J.C., GLASS, H.D. and WILLMAN, H.B. 1963 Mineralogy of glacial tills and their weathering profiles in Illinois. *Illinois Geol. Survey Circ.*, 347.

FRÖHLICH, F. 1960 Beitrag zur Geochemie des Chroms. *Geochim. et Cosmochim. Acta*, **20**, 215-240

GLENTWORTH, R., MITCHELL, W.A. and MITCHELL, B.D. 1964 The red glacial drift deposits of north-east Scotland. *Clay Minerals Bull.*, **5**, 373-381.

GOLDBERG, E.D. 1954 Marine geochemistry I. Chemical scavengers of the sea. *J. Geol.*, **62**, 249-265.

GOLDBERG, E.D. and ARRHENIUS, G.O.S. 1958 Chemistry of Pacific pelagic sediments. *Geochim. et Cosmochim. Acta*, **13**, 153-212.

GOLDSCHMIDT, V.M. 1954 *Geochemistry.* Oxford Univ. Press, 730 pp.

GOODYEAR, J. 1962 X-ray examination of some east Yorkshire boulder clays. *Clay Minerals Bull.*, **5**, 43-44.

GRAF, D.L. 1960 Geochemistry of carbonate sediments and sedimentary carbonate rocks. Part 3, Minor element distribution. *Illinois Geol. Survey Circ.*, 301.

GRIFFIN, J.J. WINDOM, H. and GOLDBERG, E.D. 1968 The distribution of clay minerals in the World Ocean. *Deep-Sea Res.*, **15**, 433-459.

GRIM, R.E. and ROWLAND, R.A. 1942 Differential thermal analysis of clay minerals and other hydrous materials. *Amer. Mineral.*, **27**, 746-761.

HARDER, H. 1961 Einbau von Bor in detritische Tonminerale. *Geochim. et Cosmochim. Acta*, **21**, 284-294.

HIRST, D.M. 1962 Geochemistry of modern sediments from the Gulf of Paria, parts I and II. *Geochim. et Cosmochim. Acta*, **26**, 309-334 and 1147-1187

HURLEY, P.M., HEEZEN, B.C., PINSON, W.H. and FAIRBAIRN H.W. 1963 K-Ar age values in pelagic sediments of the North Atlantic. *Geochim. et Cosmochim. Acta.* **27**, 393-399.

LEVINSON, A.A. and LUDWICK, J.C. 1966 Speculation on the incorporation of boron into argillaceous sediments. *Geochim. et Cosmochim. Acta*, **30**, 855-861.

MACDOUGALL, J.D. and HARRISS, R.C. 1969 The geochemistry of an Arctic watershed, *Can. J. Earth Sci.*, **6**, 305-315.

MANHEIM, F.T. 1965 Manganese-iron accumulations in the shallow marine environment. In. *Symposium on Marine Geochemistry*, Univ of Rhode Island Occasional Publ. **3**, 217-275.

MCMURCHY, R.C. 1934 Structure of chlorites. *Z. Kristallogr.*, **88**, 420-432.

MOORE, J.R. 1963 Bottom sediment studies, Buzzards Bay, Massachusetts. *J. Sediment Petrol.*, **33**, 511-558.

——— 1968 Recent sedimentation in northern Cardigan Bay Wales. *Bull. Br. Mus. nat. Hist.* (Miner) **2** No. 2, 131 pp.

MÜLLER, G. 1967 Strontium distribution in recent Indian Ocean sediments off the eastern coast of Somalia. *J. Sediment Petrol.*, **37**, 957-960.

PERRIN, R.M.S. 1957 The clay mineralogy of some tills in the Cambridge district. *Clay Minerals Bull.*, **3**, 193-205.

PETTIJOHN, F.J. 1957 *Sedimentary Rocks.* Harper, New York, 718 pp.

PHILLIPS, A.H. 1922 Analytical search for metals in Tortugas marine organisms. *Publ. Carnegie Instn.* **18**, 97.

PILKEY, O.H. and GOODELL, H.G. 1963 Trace elements in recent mollusk shells. *Limnol. Oceanogr.*, **8**, 137-148.

RANKAMA, K. and SAHAMA, T.G. 1950 *Geochemistry*, University of Chicago Press.

TAYLOR, S.R. 1964 Abundance of chemical elements in the continental crust: a new table. *Geochim. et Cosmochim. Acta*, **28**, 1273-1285.

THOMPSON, T.G. and CHOW, T.J. 1955 The strontium - calcium atom ratio in carbonate secreting marine organisms. *Papers in Marine Biology and Oceanography*, 20-39.

TUREKIAN, K.K. 1964 The marine geochemistry of strontium. *Geochim. et Cosmochim. Acta*, **28**, 1479-1496.

——— and KULP, J.L. 1956 The geochemistry of strontium. *Geochim. et Cosmochim. Acta*, **10**, 245-296.

——— and WEDEPHOL, K.H. 1961 Distribution of the elements in some major units of the Earth's crust. *Geol. Soc. America Bull.*, **72**, 175-192.

VINOGRADOV, A.P. 1953 *The Elementary Chemical Composition of Marine Organisms.* Memoir 2, Sears Foundation for Marine Research. Yale Univ. New Haven.

WANGERSKY, P.J. and JOENSUU, O. 1964 Strontium, magnesium and manganese in fossil foraminiferal carbonates. *J. Geol.*, **72**, 477-483.

WEBER, J.N. and MIDDLETON, G.V. 1961 Geochemistry of turbidites of the Normanskill and Charny formations. *Geochim. et Cosmochim. Acta*, **22**, 200-288.

WILDEMAN, T.R. 1970 The distribution of Mn^{2+} in some carbonates by electron paramagnetic resonance. *Chem. Geol.*, **5**, 167-177.

WILKLANDER, L., LOTSE, E. and VAHTRAS, K. 1956 Mineralogiska och fysikalisk - kemiska undersokningar av lerfraktionen fran odlade jordar: *Svenska Foreningen for Lerforskning*, no. 6, Okt. (Geol. Foren. Fornandl, v. 78, no. 3) p. 526 (with English abstract).

Formerly at the Institute of Geological Sciences,
now at:
Department of Geology
University of Ottawa
Ottawa 2
Canada

Dd. 502158 K 8

Printed in England for Her Majesty's Stationery Office
by Swift (P&D) Ltd., London, EC1 M 5 RE